NSF MOSAIC READER

HUMAN EVOLUTION

AVERY PUBLISHING GROUP INC.

Wayne, New Jersey

The articles contained in this volume were selected from original works that appeared in *Mosaic* magazine. They are reprinted by permission of the National Science Foundation.

Mosaic is published six times yearly as a source of information for the scientific and educational communities served by the National Science Foundation. For more information regarding *Mosaic,* please direct your inquiries to: Editor, *Mosaic,* National Science Foundation, Washington, D.C. 20550.

The publisher is indebted to Warren Kornberg, Editor of *Mosaic,* for his editorial guidance, his invaluable suggestions, and his patience. Avery also wishes to thank the members of its own editorial board for their help in article selection. Our thanks go to Paul Biersuck, Department of Biology, Nassau Community College; Mel Gorelick, Department of Biological Sciences, Queensborough Community College; John Burkart and Loretta Chiarenza, Department of Biology, State University of New York at Farmingdale; Bernard Tunik, Department of Biology, State University of New York at Stony Brook; John Maiello, Department of Biology, Rutgers— The State University of New Jersey; and Donald Wetherell, Biological Science Group, University of Connecticut at Storrs.

Cover design by Martin Hochberg.
Cover photo credit: Martin Hochberg.
In-house editor: Joanne Abrams.

Copyright © 1983 by Avery Publishing Group, Inc.

ISBN 0-89529-174-6

Printed in the United States of America

10 9 8 7 6 5 4 3 2 1

CONTENTS

Beyond the systematic study of skulls, bones, and stone tools
to trace human origin, scientists are now examining
evidence of the earth's dynamic climatic past to learn what
it was that might have contributed to the direction of
human evolution. From this study of past ecological
settings is emerging a clearer understanding of the intermix
of factors that produce biological change. Within this
framework, the author examines current theories of
evolutionary development and the pursuit of the story of
human origin.

The work of paleontologists has traditionally involved efforts
to reconstruct prehistoric life forms through estimates of
where fossil bones are found and how they fit together.
Additionally, much scholarly guesswork defined these
creatures' appearances and lifestyles. Now, new tools and
techniques are providing better information on how a fossil
segment might really have functioned and how the creature
of which it was a part might have lived and moved.
Focusing on the shape of bones, the location of muscles,
and the wear on teeth, this article examines the anatomical
work being done to delineate the course of species
evolution.

Examining some of the current theories and techniques of
paleoanthropology, this article discusses some of the factors
that appear to link *Ramapithecus*—believed by many to be
a very early member of the human family—to *Homo
sapiens sapiens*—modern humankind. Specifically, through
the analysis of adaptation to diet shifts caused by changes
in the climate, the research seeks not only to reconstruct
the appearance of this hominid candidate, but also to
determine its behavior and mode of living.

The Emergence of Homo sapiens

Did modern people descend from a line that replaced the
Neanderthal in Europe, or did humankind evolve directly
from the Neanderthal? For years, scientists have been arguing
both sides of this question. Basing the discussion on the
latest available data, this article reviews, analyzes, and
evaluates both schools of thought.

Pre-Clovis Man: Sampling the Evidence

An astounding ninety-nine percent of the Western
Hemisphere's archaeological sites have been destroyed by
climate or geological processes. Analyzing the bones and
tools that have been found, scientists have concluded that
the first inhabitants of the Western Hemisphere were
hunters on the trail of their game, not migrants seeking
new homes. Just who these earliest emigrants to the New
World were, the routes they traveled, and the way they
lived are still the subject of often heated debate among
archaeologists.

Cultural Evolution

In an effort to determine how early hominid societies
interacted among themselves and with their environment,
scientists have been observing both primitive human societies
and nonhuman social animals. Extant primitive cultures
enlighten us as to the patterns of hunter-gatherer societies.
Birds, chimpanzees, and even ant colonies are scrutinized
for clues to the evolution of group behavior. The author
presents selected studies and discusses the contributions
they have made to the understanding of human cultural
development.

INTRODUCTION

Mosaic, the source of the articles in this reader, is the bimonthly magazine of the National Science Foundation. Its purpose is to keep nonspecialists in any of the sciences aware of the ferment at the frontiers of many scientific research disciplines and the research trends out of which tomorrow's scientists will emerge.

Mosaic's purpose is to explore the thinking of researchers about both the current and future status of their science. Its articles provide insight into the problems facing investigators in virtually every research area, and explore the ways they seek to overcome those problems—often by crossing traditional disciplinary lines.

Prepared by experienced science journalists and authenticated by scientists, these articles reveal the processes of science, as well as its progress. They not only report on day-to-day advances but offer perspectives on what science is.

For the *Mosaic Reader Series,* recent issues of the magazine have been surveyed, and groups of articles assembled to provide broad, pertinent overviews of segments of scientific research. This reader on human evolution considers the recently developed tools in use and the fresh and exciting insights emerging as scientists of many disciplines explore man's past. Critical to the new view of this subject are the interrelationships among living things and living systems that are so much a part of modern thought. Environmental change that might have driven human development is being considered, as well as the impact that emerging humanity might have had on the world it grew to occupy.

There are many pieces missing from our picture of human evolution. But if scientists are in concurrence regarding anything in this complex area, it is this: they realize that it will take an array of scientific perspectives to satisfactorily answer the new questions that each new answer in this fascinating, ever-developing field seems to produce.

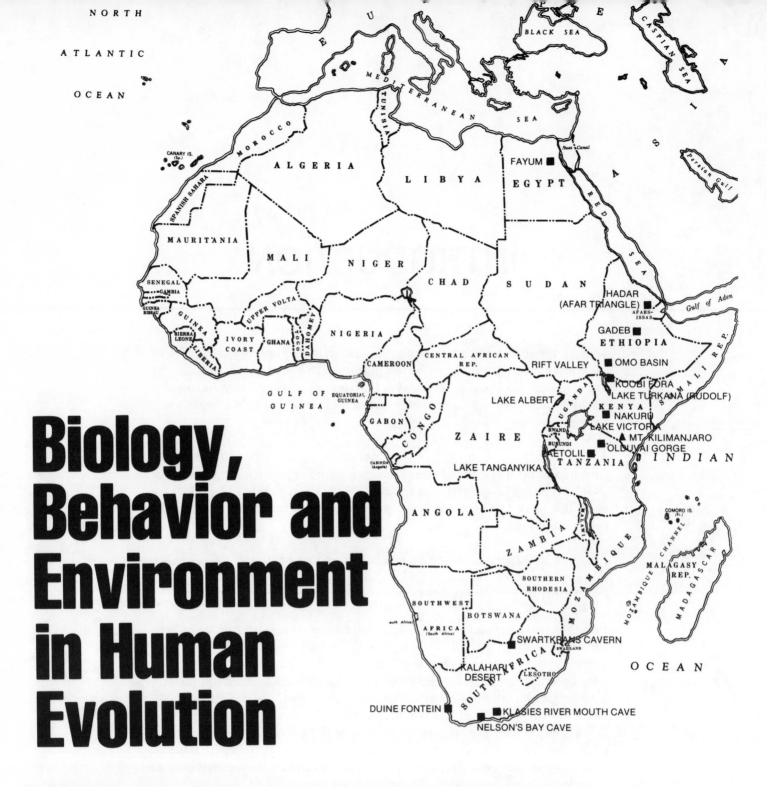

Biology, Behavior and Environment in Human Evolution

Contributions from a score of disciplines are replacing guesswork in the description of the environments that accompanied the stages of humankind.

PRIMATES ———————————→

65	60	55	50	45	40	35

MILLIONS OF YEARS

←——— PALEOCENE ———→ ←——————————— EOCENE ———————————→ O

←———————————————————————————————— TERTIARY ———————————————→

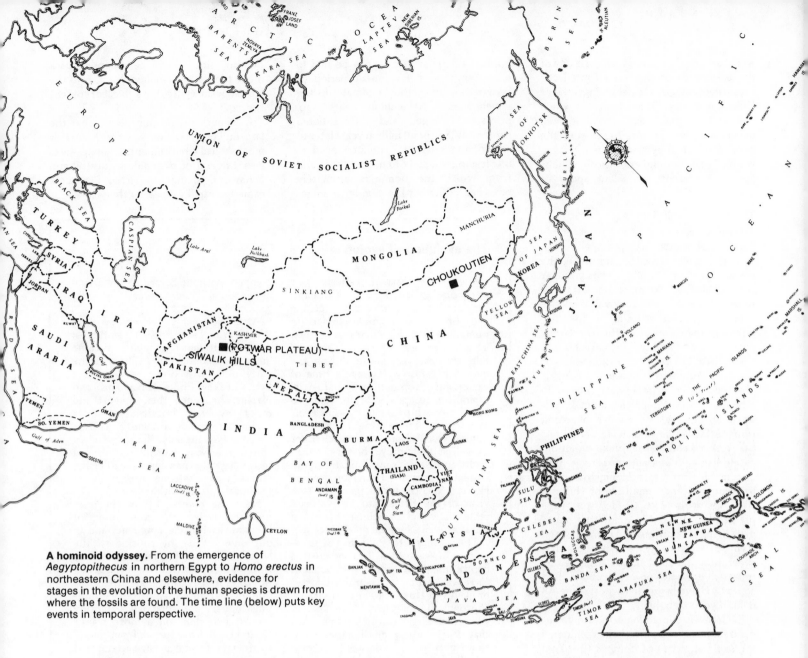

A hominoid odyssey. From the emergence of *Aegyptopithecus* in northern Egypt to *Homo erectus* in northeastern China and elsewhere, evidence for stages in the evolution of the human species is drawn from where the fossils are found. The time line (below) puts key events in temporal perspective.

An estimated 130 million people, the largest audience in history, watched the television presentation "Roots" two years ago. One black American's search for his family origins in Africa touched a universal human chord—curiosity about our past. In its response to "Roots," despite subsequent controversy regarding the research, the public's fascination transcended race and time to share the success of author Alex

Haley in tracing his ancestry to a West African youth named Kunta Kinte, who was captured into slavery two centuries ago.

Between that "noble savage" and his great-great-great-great grandson stretch seven generations of time and a world of difference. Yet we have a faint "feel" for that 18th century in which the enslaved Kunta Kinte grew to manhood. It lies on the far side of what primatologist

Sherwood Washburn of the University of California at Berkeley and others call "social time," just beyond the familiar "biological time" of generations alive today but this side of both "historical" and "geological" time, which extend back to the limits of the ability of a human brain to comprehend, and beyond.

Nevertheless, people share an intense curiosity about their uttermost roots. "The satisfaction of knowing how the

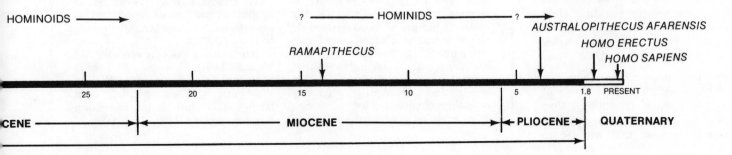

world assumed its present shape is one of the basic ends of the human intellect," says paleoecologist Daniel Livingstone of Duke University. "It is a thing for which we strive from earliest childhood, questioning our parents for tales of the days in which they and the world were young." To youngsters, those distant epochs represent equivalent spans of time.

The play of evolution

Without recourse to the roots of life itself, the events that presage human evolution can be traced to a time when the world was 30 or 40 million years younger than it is today. Like most times, it was a time of change. "The survival of species depends on their adapting to the conditions of their time...[on] the fit between population and reality," says Washburn.

For "reality" read "environment." "If the environment doesn't change," observes Duke primatologist Elwyn L. Simons, "neither will the animal." Ultimately, it was the earth's changing climate, landscape and living conditions that sparked and fueled human evolution.

"Humanity responds not to a unidirectional shift," says Livingstone, "but to frequency and amplitude of shifts in the environment." He portrays prehistoric humans as mid-sized mammals living by hunting and gathering and trying to adjust to swings in temperature, vegetation and food supply triggered by repeated episodes of cold climate that appear to have begun in earnest some 1.8 million years ago.

Such severe climatological and environmental swings appear to have dominated the last half or so of hominid (mankind and its unique, linear, non-ape antecedents) evolution. Hominid evolution goes back at least to the earliest known remains of an omnivorous, bipedal creature called *Australopithecus*, dating back some four million years in Africa and disappearing in the face of more advanced competition only some million years ago.

As much as the last million years of *Australopithecus* could have overlapped the time on earth of *Homo erectus*, the advanced tool- and culture-maker that led some 100,000 years ago to the first *Homo sapiens*—*Homo sapiens neanderthalensis*—which gave way to modern *Homo sapiens sapiens* only within the last 40,000 years.

Those last two million years, of *Australopithecus* and *Homo*, were a time of accelerating hominid evolution. They witnessed the vastly increased pace of the evolutionary process that had led to humankind. That process can be said to have begun in a far longer period of benign and less variable climate during which, some 90 million years ago, primates emerged and differentiated some 50 million or 60 million years later to spawn homin*oids* (the common primate line leading uniquely to man *and* apes and from which the hominids ultimately branched). Whether the uniquely human, hominid line should be extended back more than 4 or 5 million years, into the late Miocene epoch, is a subject of current controversy.

It is, perhaps, no coincidence that the last million years or so—the period of most radical environmental change—was also a period of pyramiding cultural and behavioral change in *Homo*. Pressures came too rapidly on each other's heels for

The evolution of evolution

Ever since Darwin, reflects Yale anthropologist David Pilbeam, Western scholars have been seeking and finding their own image in humankind's evolutionary ancestors. This "ideology of anthropology," Pilbeam observes, began during the mid-19th century. It was "a time of expanding competitive industrial capitalism, so apes were presumed to have become humanized by using tools as weapons." These upright killer-apes hunted, clubbed and fought each other as well as their prey. "Nature, red in tooth and claw," sang England's poet laureate, Alfred Tennyson, while Herbert Spencer proclaimed "the survival of the fittest" to be Nature's pitiless law and Darwinism became a social as well as a biological precept.

Then, suggests Pilbeam, as confidence in Western civilization crested before the first world war, ancient *Homo* was cast as "a big-brained, lofty-browed creature of lofty morals." (Apropos the morality, Duke University primatologist Elwyn Simons notes that 19th-century missionaries, who shot gorillas in "darkest" Africa, described these presumed precursors of humankind as leading strictly monogamous lives—which, Simons says, "was and is not at all true.") And Neanderthal Man, whose skull had first surfaced in the mid-eighteen-hundreds, was depicted as apelike, beetle-browed, bowlegged and brutish, to prove him an early—hence coarse and unrefined—forerunner of perfected, highbrow man.

(Berkeley anthropologist Sherwood Washburn recalls that when Peking Man came to light in the nineteen-twenties, "he was regarded [in the West] as too primitive to have made stone tools, although tools were found alongside his bones." Further, Washburn notes, assuming tacitly that humankind's emergence was an onward-and-upward progression, early anthropologists tended to arrange stone tools "in orders from simple to complex without regard to the actual associations and orders in the ground.")

After advancing technology won World War Two, Pilbeam continues, "the evolutionary metaphor became 'man the toolmaker.'" Thus in the early nineteen-sixties, fossil-hunters Louis Leakey, Philip Tobias and John Napier christened their newest and most evolved hominid find, a toolmaker found at Olduvai Gorge, *Homo habilis*—literally, "handyman." In that postwar period of rapid technological expansion, says Pilbeam, "there was a strong synonymy of technical culture with culture."

Next, he points out, in the later nineteen-sixties, the touchstone of humanness became language, and in our own decade, he concludes, "the latest intellectual fashion or discovery is to see complementarity of the sexes, 'sharing and caring,' as the basic human attribute." And today's multidisciplinary anthropology, exploring human evolution on a more fragile "island earth," has turned from "links" to lifeways, from exclusive attempts at profiling ancient man to reconstructing the environments and ecologies in which he evolved.

To conceive of man as evolving from a rough brutal intermediate form to the refined finished product "is a myth," declares Kay Behrensmeyer of Yale. Each hominid stage in its own time and place, she declares, was thoroughly "evolved" and sophisticated to fill its ecological niche, rather than existing in a holding pattern on the way to becoming *Homo sapiens*. Or as David Pilbeam puts it, "We can look at evolution now as an opportunistic process; it was not set in motion simply to end in us!"

"The missing link between apes and human beings," wrote a French philosopher, "is man." Writes Washburn: "When people, both laymen and scientists, wanted missing links, they made them."•

biological adaptation in so long-lived a creature, Duke's Dan Livingstone offers, "but, being human, they could adapt culturally."

So overly linear, broad-brushed a picture, of course, obscures many details. Karl W. Butzer of the University of Chicago, for instance, repeatedly uses the word "mosaic" to describe the complex ecological settings in which primate, hominoid and hominid evolution took place across tens of millions of years, along with a concomitant "mosaic evolution and ecological speciation...."

Nevertheless, how cultural prowess, broadly, grew out of and eventually overgrew the anatomical and behavioral leaps that turned hominoids into hominids and ultimately into humans, what mix and interplay of environmental change and biological and behavioral evolution contributed to the emergence of humankind, has become the focal riddle of modern paleoanthropology.

Until only 10 or 15 years ago, the scientists who sought systematically to trace humanity's origins were essentially fossil-hunters, sorting skulls, bones and stone tools into postulated sequences of age and sophistication. With their attention focused on humanity's family tree, they often paid little heed to the forest in which it grew.

That's all changing now. Paleo-anthropology today is part of an intricate matrix of such sciences as paleoecology, paleoclimatology, palynology, geology, geophysics and taphonomy. (Palynology is the branch of paleobotany that reconstructs bygone landscapes through analysis of fossil pollen; taphonomy is a study of dynamic processes undergone by bones between the time they are laid down and the time they are recovered.)

"You can't understand human behavior if you don't have it in its natural setting," says Berkeley anthropologist F. Clark Howell. Agreeing, David Pilbeam, curator of the Yale-Peabody Museum, sees himself and his fellow anthropologists as stage hands "reconstructing scene changes in the ecological theater where the play of evolution is performed."

To set the stage

A curtain-raiser to the drama of human origins took place some 40 million years ago when the subcontinent of India, drifting north on its tectonic plate, fetched up against the mainland of Asia. Before that, writes Karl Butzer, had been "a time of warm and little-differentiated climate...subtropical or semitropical environments prevailed in middle

Earliest bipedalism. Footprints of two upright, walking individuals, dated to some 3.7 million years ago, are the earliest known evidence of bipedal gait in hominoids. Found by Mary Leakey and others, the footprints in volcanic tuff are filled with black sand for contrast.

National Geographic Society

latitudes...even the high Arctic was temperate...modern summer weather patterns prevailed even in the winter hemispheres...."

But the tectonic crunch that forced the Himalayan Mountains to rise along the line of impact created an atmospheric high dam that deflected the earth's prevailing wind patterns, and global temperatures dropped. Additionally, the larger, Antarctic land mass shifted fully into polar position, changing circulation patterns in the southern ocean and contributing to worldwide climate change.

Then, some 18 million years ago, continued tectonic drift drove the Afro-Arabian land mass into contact with southern Europe and southwest Asia. Animals which had been evolving in changing African habitats could now spread into Eurasia over dry land.

Among the migrants may have been humanity's emerging ancestors. "Our unclothed thermal tolerance and our ability to synthesize vitamin D except in sunlight...suggest that we are primevally creatures of a tropical environment," offers Livingstone. But we are not creatures of the lush, steaming stagnation of wet, tropic climes, he notes; such an environment would not goad so dramatic an evolutionary process. Except for a narrow belt of hothouse jungle along the Equator, Africa is—and for most of prehistory has been—desert and semi-arid grassland, parkland, steppe and savanna, with alternating wet and dry seasons. Rooted in such a setting, primordially, "man is a savanna animal," Livingstone observes, as he was when the stage on which the drama of human origins was to be played spread from one continent to three.

A radiation of apes

Evidence of the earliest hominoids we know, primates—the characteristics of which hint at future hominid traits—comes to us from Africa's far northeastern corner, not far from the 18-million-year-old Afro-Eurasian junction. The Fayum Depression there is a desolate stretch of torrid, bone-dry badland just west of the Nile Valley on the edge of Egypt's Libyan Desert. Its fossil-bearing sites are about 65 kilometers from Cairo and 40 kilometers from the nearest human habitation. Remains of its hominoid inhabitants have been found in rocks inferred to be as much as 30 million years old. It is a paleoecological setting of a comprehensible, less-than-global size.

"The evolution of tertiary higher primates," Butzer observes, can hardly be discussed in terms of generalities at a global scale. "Instead, different primates probably favored different meso- or microenvironments within comparable macroenvironments. It is, therefore, at the local habitat level that paleoecological studies must be focused."

In such a tract of the Fayum, measuring some 3 kilometers by 13, are concentrated 95 percent of the Oligocene traces of land mammals so far found in Africa. These deposits span roughly the time between 40 million and 29 million years ago. And there, since 1961, frequent expeditions led by Elwyn Simons have found more than 400 fossils of as many as 11 kinds of primitive, monkey- and apelike, four-footed, tree-dwelling primates. Of these, about 40 specimens represent half a dozen species of ape.

"Success in fossil-hunting calls on four factors," Simons says, "motivation, experience, keen eyesight and brains." (Luck is "equally important," comments a colleague.) Simons was deploying all four (or five) in 1963 when, while combing the windswept, rock-strewn surface of the Fayum beds, he found an inch-long jawbone fragment that may be 29 or 30 million years old. Simons and his Fayum fossil hunters subsequently found a complete skull and several forelimb bone fragments. In 1965, he named the creature that bore them *Aegyptopithecus*—"ape-of-Egypt."

The largest primate of its day, *Aegyptopithecus* "is a broad approximation," Simons suggests, "of what an ancestor of man at that time should be like." Judging by its apelike teeth, this cat-sized hominoid fed on fruits and leaves; it was probably "a generalized, arboreal quadruped with arm bones that suggest the modern South American howling monkey" more than any other living primate, says Simons. But, he adds, "we don't know and can never know if it was truly a founding father of the human race."

Out of the trees

In the now treeless, lifeless Fayum there are petrified tree trunks more than 30 meters long. Near them have been found seed pods identical to those of African plants that grow today in standing water. They testify to the evergreen forests that lined multichannel riverbanks in ancient Fayum times; their verdant branches were home to quadrupedal, arboreal leaf browsers and fruit eaters the likes of *Aegyptopithecus*. As the lemurs in Elwyn Simons's primate center at Duke do to this day, they probably slaked their thirst by licking raindrops off the leaves.

There is no inference, Simons says, of climate change in the Fayum of that time. Nevertheless, there is evidence that, during that time, the Fayum monkeys were venturing out of the trees to drink at the stream. Simons deduces this from the tender age of so many of the monkeys, the fossil remains of which he finds in the sandstone sediment of former riverbeds. These thirsty, clumsy young ones, he speculates, must have been picked off the bank by predators—turtles, crocodiles and the like—as they approached the streams to drink.

Later changes in tooth form documented in the fossil record indicate that, increasingly, the hominoids were eating as well as drinking on the ground, at least part of the time, as climatic and environmental change ultimately imposed new pressures on them. They grew in bulk; their molars evolved in size and thickness of enamel to masticate the fibrous roots and grasses and crack the hard-shelled nuts and seeds of the wooded African parkland and savanna that was patchily replacing the jungle.

Though eastern and southern Africa are certainly where most of the later—four million years and younger—hominid remains have been found, it is evidence from more northern climes that links them over time to their hominoid antecedents. Africa may not have been harsh enough to drive the engine of change in the direction of hominid evolution. Simons suggests that "the birthplace of the primate branch that led to man was along the northern edge of the distribution of the larger Miocene apes of Europe and Asia. That's where the environmental challenge was."

As apes "radiated" and spread north from Africa some 17 to 15 million years ago, when the climate and landscape changed, "they found it harder to make a living.... Unlike the tropical forests, where trees bear fruit the year around," Simons observes, "the temperate-zone plants yielded only seasonally; the big newcomers had to forage on the ground for leaves, bark, roots, grass and nuts." We know they were ground dwellers in bush country, Simons points out, not only because of their size, but also because their bones typically occur together with those of grazing animals.

Most of the European and Asian Miocene apes, however, were apparently not destined to make the leap from the hominoid in the direction of the hominid. A candidate for that distinction—if indeed

it happened only once—Simons and others identify as a collateral offshoot of those earlier African apes—a smaller perhaps more versatile creature named *Ramapithecus*, whose fossil jaws and teeth have been found in northern India, Pakistan, the Balkans and East Africa and who—so far—occupies center stage in Act Two of the drama.

Reconstructing the ecosystem

The setting for Act Two shifts from Africa to the Siwalik Hills of northern Pakistan, at the foot of the Himalayas. The rapidly eroding rock formations there expose former land surfaces one to 14 million years old. The rugged Siwalik Hills have been a fossil hunter's happy hunting ground for more than a century. Among 13,000 teeth and bones so far collected in an area there no bigger than Rhode Island, 86 are specimens of hominoid primates from at least 43 individuals, including *Ramapithecus* and several of his bigger cousins.

The claim of *Ramapithecus* to a hominid place at the root of humankind's family tree rests so far on evidence adduced from his big cheek teeth, thick tooth enamel and remarkably small canine teeth, as well, Simons says, as numerous details of mandibular and facial architecture in the fossil record. *Ramapithecus* is unique among late Miocene, quadrupedal apes, Simons contends, in the delayed eruption of third molars—literally "wisdom teeth"—in some specimens. This, Simons points out, suggests a long ramapithecine childhood, a growing-up time in which to learn behaviors that might be associated with group-living characteristics beyond those of other hominoids.

"But a skull or a jaw tells only a fraction of the story," demurs David Pilbeam, who has led annual expeditions to Pakistan since 1973. To identify hominid traits, "We need to reconstruct the complete picture of how a hominoid lived—as an animal."

Even if *Ramapithecus* was the first recognizable hominid on earth, Pilbeam emphasizes, he was not the only hominoid in the complex mosaic of grassland and forest habitats in Pakistan. Another, or others, might have made the grade.

The three or four other and larger apelike species trying to make a living down from the trees between 14 and 8 million years ago all had the big cheek teeth and thick enamel of ground foragers. "It looks as if the hominoids were experimenting with living outside the forest and away from the trees, some of the time on the ground," Pilbeam speculates, "experimenting with being omnivorous," enjoying the best of both worlds, the arboreal and the terrestrial.

In our more immediate ancestors, the hominid line, he adds, "early omnivory is important in setting the scene for culture." In Pilbeam's tentative version of human evolution, omnivory did not develop gradually as an adaptation to habitat, but may well have been provoked rapidly in response to a more dynamic ecological challenge.

The missing chain

But what that challenge was—what actually happened between the end of

Settings. Savanna landscape in Tanzania (left), typical of mixed tree and grassland environments in which human evolution took off. Siwalik Hills in Pakistan (above), happy hunting ground for ramapithecine fossils. Horse and buffalo jawbones at a Stone Age butchery site (bottom) at Duine Fontein, north of Cape Town.

Dan Livingstone; David Pilbeam; Richard Klein

ramapithecine Act Two and the beginning of australopithecine Act Three—is unknown. Unfortunately, our view of *Ramapithecus* and its Miocene cousins grows dim toward the end of Act Two. And many of the early Act Three pages are missing. Whether his line continues, to join that of *Australopithecus* and, ultimately, *Homo*, is the question that is increasingly being asked—not only of the fossil record, but of the paleoenvironment as well.

For *Ramapithecus* and its hulking

hominoid relatives vanished from the stage of prehistory as far as available evidence is concerned. No unequivocal fossil trace of possible human precursors has been found anywhere in the world until *Australopithecus*, a full-fledged, upright, omnivorous hominid, burst on the African anthropological scene four million years ago. This intermission after Act Two would have begun eight or nine million years ago and lasted through the late Miocene epoch, a time of fringing forests and intertwined landscapes of open woodland and grassland.

In such a dynamic environment, mammalian evolution moved fast, but for hominoid remains, that period of four or five million years—except for two or three equivocal fragments in Kenya—is a fossil void, a blank.

Elwyn Simons was among the first to point out this gap in the prehuman record between hominoid and hominid. To Clark Howell of Berkeley, "It's not just a 'missing link'; it's a whole missing chain, a big black hole!" Unless and until many more hominid fossils are recovered from this gap period, says Pilbeam, "it's really not possible to say that *Ramapithecus* is indeed ancestral to later hominids."

A giant step

The Act Three curtain rises, well into the Pliocene, almost four million years ago, on a big hominid and a small hominid walking side by side across a barren plain. The rainy season was starting in the high, dry grasslands, and a light drizzle bespattered the level layer of fine volcanic ash several centimeters deep that had just buried the landscape.

As the pair of upright, bipedal primates plodded over the dampened plain, they were not alone. The land was alive with moving beasts, from elephants to three-toed horses to millipedes. Each pressed its peculiar track into the soft, raindrop-indented ash.

Today, almost four million years later, that smooth surface "is as hard as a concrete freeway," says physical anthropologist Tim D. White of the University of California at Berkeley. The

---

Fossil hunters. Scientists in the search for human origins include (clockwise from top left) Karl Butzer at Swartkrans cavern; Kay Behrensmeyer at Koobi Fora; Dan Livingstone preparing to core in Lake Chishi, Zambia, in 1960; Elwyn Simons and the jaw of *Aegyptopithecus*; David Pilbeam in the Siwalik Hills; Garniss H. Curtis (left) and Robert Drake potassium-argon dating Laetolil tuff; Donald Johanson (left) and Tim White, who named *Australopithecus afarensis*.

---

site, known as Laetolil, lies in northern Tanzania. It is some 50 kilometers south of the Olduvai Gorge made famous by Louis and Mary Leakey's quest since 1931 for a cradle of ancient man.

The fossil hominid footprints discovered since 1977 by Mary Leakey and her colleagues at Laetolil "push the evidence of bipedal locomotion back to almost 3.8 million years," says Tim White, who, with Yves Coppens of Paris was among the discoverers. The big toes pointed straight ahead in the manner of human feet, rather than projecting sideways, as do apes'.

Additionally, not far from the footprint layer in the Laetolil tuff, or petrified ash, native Kenyan fossil prospectors on Mary Leakey's team (supported largely by the National Geographic Society) have since 1974 found two-dozen jaw and tooth specimens. They are clearly of manlike proportions, though extremely primitive. This bone-bearing rock has been firmly dated to between 3.6 and 3.8 million years ago, just when those footprints were made. It pushes back the date at which a primitive human ancestor walked erect.

Mention of prints in volcanic ash lasting for millions of years makes Milford Wolpoff of the University of Michigan sigh: "Why couldn't one of the little fellows have stumbled and fallen down, so we'd know what he looked like?"

Those teeth and jaws discovered in the Laetolil tuff are reminiscent of humankind's shadowy Miocene ancestor, back across that multi-million-year gap. And, says Tim White, they are indistinguishable from those of other, younger and more complete fossil skeletons unearthed 1,600 kilometers to the northeast by Donald C. Johanson, curator of physical anthropology at the Cleveland Museum of Natural History. At Hadar, a site in Ethiopia's Afar triangle, Johanson's annual expeditions have dug up hundreds of fossil hominid bones representing between 35 and 65 individuals that lived between 2.6 and 3.3 million years ago. The best known of these is "Lucy," a petite and presumably female creature, 40 percent of whose skeleton has been reassembled. Limb and pelvic bones argue that the Afar hominids, like those at Laetolil, walked on two feet.

Although these remains are half a million years younger than the Laetolil fossil footprints, the two are so similar that Johanson and White have agreed to christen their protohuman finds with the same name—*Australopithecus afarensis*. It is, in their words, "the most primitive group

of demonstrable hominids yet recovered from the fossil record and allows a perception of human evolution hitherto impossible."

From this side of the four-million-year gap since *Ramapithecus* dropped out of sight, Johanson says of *Australopithecus afarensis*, "While clearly hominid in their dentition, mandibles, cranium and post-cranium [face, body and limb bones], these forms retain hints of a still poorly known Miocene ancestor." Also, both Simons and Pilbeam, he notes, have seen in Lucy's jaw "interesting things that are reminiscent of *Ramapithecus*."

All of the hominids whose bones have been found in the Afar triangle, Johanson emphasizes, "were walking bipedally three million years ago just as well as you and I walk to and from lunch." This means, he explains, that bipedalism, along with omnivory, a *sine qua non* of emergent humanness, "may have quite a longer ancestry than we thought before."

In this, Berkeley's Sherwood Washburn finds support for his argument that "bipedal locomotion is not just another human anatomical adaptation, but the most fundamental one," presumably going back four million years or more.

This suggests, however, that evolution "dragged its feet" for two million years after straightening man up onto two legs. The first stone tools, showing that he could use his head and hands as well as his feet, don't appear until two million years later and *Homo erectus*, with his big brain, a million years after that. (Agriculture, one cultural payout of braininess, did not occur to bipedal, stone-flaking, hunting-and-gathering humans until a mere 10,000 years ago, some 30,000 years after neanderthals were succeeded by modern man.)

Reading the pollens

Africa sheds its sediments toward the coasts, explains Elwyn Simons. This is why so many fossil-bearing deposits are found in the great Rift Valley that cleaves its eastern side. As at Laetolil and Hadar, this highly volcanic seismic scar has buried successive ages of bones and footprints in layer after layer of compacted ash. Thus, many multidisciplinary multinational teams of scientist are now working in several sites along the rift to find the bones, stones and environment of ancient man.

One day, 2.8 million years ago, lava flowed down from a nearby volcano and dammed the Webi Shebele Valley in central Ethiopia's Plain of Gadeb. A lake formed, more than 30 kilometers long by

15 kilometers wide. This dam gradually gave way between 1.5 million and 700,000 years ago, and *Homo erectus* camped on the former lake bottom, leaving stone tools in abundance for J. Desmond Clark of Berkeley to recover. *H. erectus*, including "Java man," "Heidelberg man" and "Peking man," probably evolved from *H. habilis*, the first toolmaker.

For five years, Clark has been exploring the Plain of Gadeb. One day in 1977, he and French paleobotanist Raymonde Bonnefille climbed down into the 300-meter-deep gorge of the river at its western end. Their quick reconnoiter of datable rock formations, from 2.7 million years old at the top to 16.5 million near the bottom, led Clark to plan intensive prospecting of the gorge in 1979. "It is possible," says the Berkeley archaeologist, "that fossil-bearing sediments may be present dating to the time of the 'missing millennia' between four and ten million years ago." And, he adds, "if this intuition proves correct, it may ultimately be possible to clarify the paleoenvironmental status of the crucial gap between the youngest currently known *Ramapithecus*-yielding rocks so far discovered in Africa."

Bonnefille, like Livingstone of Duke, is a pioneer in the pollen-tracking discipline of palynology. At her laboratory in the French National Research Center in Marseilles, she is reconstructing the history of vegetation around ancient Gadeb Lake from the fossil pollen she has collected. "It is the richest site in Africa," she says. This kind of environmental data will be correlated with the varied types of crude and advanced stone tools Clark is finding in the area.

A Noah's ark

Another rift lake, though not a fossil one like Lake Gadeb, lies near the corner where Ethiopia, Kenya and Sudan come together. There, the Omo River flows south through Ethiopia into Lake Rudolf. All but its northern shore lies in Kenya, which has rechristened this 250-kilometer-long body of water Lake Turkana.

From 1966 to 1974, anthropologists F. Clark Howell of Berkeley and Yves Coppens of France have led annual expeditions to the Omo basin. In 1978, extensive paleomagnetic dating studies of its fossil-bearing rocks were completed. They span more than two million continuous years, from 3.3 to 1.0 million years ago. Some 260 hominid teeth and bones collected at the Omo site over an area of as many square kilometers, covering some two million years of time,

range from *Australopithecus* to *Homo erectus*.

Howell is not just digging up fossil forebears; he and three-dozen specialists are sifting, literally, the evidence of past plant and animal life to put ancient hominids in their place ecologically and so to see how they came to be human. In the effort, Howell has amassed over 45,000 fossil specimens of ancient fauna—a vast zoo of bones from fish and shellfish, crocodiles and elephants, rhinos, horses, antelope, pigs, hippos, saber-toothed tigers, hyenas, cheetahs, giraffes, baboons, monkeys, apes, bats, rats and mice.

This Noah's ark of primeval beasts in the hands of computer-aided paleo-ecologists is helping to recreate the climate and landscape of the Omo basin in which its humans evolved. "Rodents, for example," Howell explains, "are very important habitat indicators, because they have large litters and change quickly in response to changing circumstances."

His field workers scoop up tons of sedimentary dirt, wash it in tons of water to float away clay and silt, then soak the residue in kerosene to help sort the coarse from the fine particles. Then, like panners for gold dust, workers with fine eyes and fine tweezers pick out the microfaunal fossils—such as shrew skulls or pinhead-sized bats' teeth. Tallying their total numbers per geologic time period fills in the environmental picture: Kangaroo rats flourished in dry open habitats, bats in moisture-loving forests; tribes of antelope ancestral to modern impala, kudu and gnu lived beside water in open grassland; some kinds of extinct pig preferred closed, wooded living space.

Climatic changes, along with geological upheavals, control shifts in the environment. Palynologist Raymonde Bonnefille can tell how the Omo basin climate has fluctuated over the past three million years. Today, she reports, the Omo is a semi-desertic, open-vegetation scrub savanna. But before *Australopithecus* roamed this earth, it was a tropical rain forest.

"None of this is really traditional anthropology," allows Clark Howell. "But it all relates to human origins."

Thus the oldest hominid artifacts in the Omo date back two million years. They coincide with the appearance in the fossil record of *Homo habilis*, the early human toolmaker, as well as with an interval in which Omo vegetation and fauna changed. Curiously, contemporary with these fine-boned or "gracile" *Homo* remains, Howell has found fossil remnants of a much more robust australopithecine hominid, which apparently did not die out

until a million years ago, after *Homo erectus* appears on the scene.

"If you had been walking around in that time," says Kay Behrensmeyer of Yale, "you would not have felt strange. Trees, river channels, monkeys, antelope of many kinds and hominids of two kinds were all there in one ecological community." She believes that the hefty *Australopithecus* and gracile *Homo erectus* coexisted in adjacent ecological niches for perhaps a million years.

It was at Koobi Fora last summer that Behrensmeyer discovered the first fossil footprints in that area. She found them in a volcanic layer adjacent to the "KB site tuff," an artifact-rich stratum she discovered in 1969 and which is identified with her initials. The three imprints, left by the feet of a bipedal hominid 1.5 million years ago, are about the size of a modern foot and could have come from *Australopithecus* or *Homo erectus*, she thinks. She expects to lay bare more of these footprints this summer and is eager to compare them with the Laetolil tracks, which are more than twice as old.

Cultural adaptations

It evidently took *Australopithecus* nearly two million years to graduate from early bipedalism to elementary stone tool making. This giant step for mankind more or less coincided with the onset of the Ice Ages in northern latitudes 1.8 million years ago. The alternating glacial and interglacial periods affected eastern Africa's climate and environment, which in turn should have accelerated hominid acquisition of human attributes.

Act Four in the drama of human evolution finds *Homo—Homo erectus—* moving from the wings to center stage, during the past million years or so, and interacting with the entrances and exits of the Ice Ages.

The million years preceding the last (65,000 to 10,000 years ago) glacial period, says Chicago's Butzer, a period marked by ten or more cycles of Northern Hemisphere glaciation that began about 800,000 years ago, "saw the disappearance of the very last australopithecines, the evolution of *Homo erectus* and the first appearance of *Homo sapiens*. This same span of time also saw a very slow development of cultural capacities that ultimately made possible the human achievements first dramatically verified in Ice Age art...." *Homo erectus*, Butzer observes, was "culturally distinct from the earlier australopithecines and the earliest bearer of human culture."

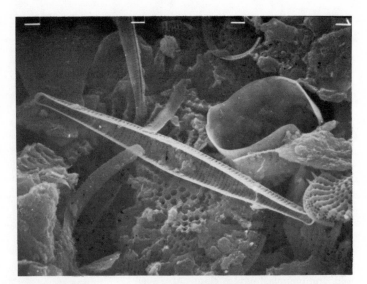

We don't know the beginnings of *H. erectus*'s cultural adaptations—changes in the way he organized his life to complement his changing innate abilities. What we do know is that by half a million years ago *Homo erectus* was using his evolved brain to devise uniquely cultural solutions to the array of challenges posed by swings in climate that deposited him ultimately in a northern, nontropical environment. Sites as dispersed as Choukoutien in China and Toralba in Spain convey a sense of *H. erectus*'s adaptations. Bones of ibex at Choukoutien, for instance, are powerful testimony to sophisticated and organized hunting skills. Fire, known by some 200,000 years ago, was protection against the harsh, glacial, mid-Asian winter. And there is evidence at Toralba of concerted game drives in which mammoths were ensnared in bogs for killing and butchering.

The record in the lakes

The ice ages were a common factor linking *Homo erectus*, through his Neanderthal descendants, to modern man. Through that time, the closer we get to the present, the finer is the detail we can find in the environmental record. In the effort to fill blanks in the record for the last 40,000 years, paleoecologist Dan Livingstone has over the past two decades sunk cylinders into the floors of a score of lakes all over Africa to recover microfossils of bygone plant life.

Analysis of these lake cores, which average 30 meters in length, has dispelled the notion that the ice ages were times of high humidity as well as low temperature. Quite the contrary, radiocarbon dating of the fossil vegetation in Livingstone's short cores from Lake Tanganyika, Lake Victoria and Lake Albert, plus botanical identification of their pollen and diatoms,

Fossil diatoms. Early holocene (10,000 carbon-14 years old) fossils cored by Joe Richardson and Bob Kendall at 14.1 meters in Pilkington Bay, Lake Victoria, in the search for ecological settings that accompanied evolutionary development. (3,500X magnification)

Dan Livingstone

---

have shown that glaciation was marked by extreme drought and cold—a dry chill climate favoring open steppe over closed forest.

For ten million years and more, recounts the Duke University palynologist, the waters of Africa's lakes have received from the atmosphere a steady downpour of pollen from trees and grasses, flowers and shrubs. These minute motes of organic dust drifted to the lake floors, accompanied by the chalky, microscopic shells of dead diatoms. With no oxygen in those deeps to decompose it, the organic sediment piled up—1 or 2 millimeters a year, 10 to 20 centimeters each century, up to 2 meters in a millennium. In a cross section of this sediment, separate plankton blooms can often be distinguished, says Livingstone, "and possibly events separated in time by as little as two weeks."

He and geologist Neil Opdyke of the Lamont-Doherty Geological Observatory in Palisades, New York, propose to drive a ten-centimeter well casing down 1,000 meters through the compacted bottom ooze of African lakes to bring up cores of mud containing a complete record of the regions' climate and ecology through all the millennia of human, post-australopithecine evolution in Africa, now hidden like a lost tape recording. They estimate that "data obtained would significantly increase our understanding of human evolution" for the million years since *Homo erectus* came on the scene.

How *Homo* got to be *sapiens*

Man's final curtain call, as fully evolved *Homo sapiens*—first as *neanderthalensis* and then as *sapiens*—finds him braving the cold dry grasslands all over the Old World, from the creeping glacial ice fronts of Europe and Siberia to the southernmost tip of Africa. As hunter-gatherer, he still had a lot to learn.

Studying how man's subsistence behavior developed in the last 130,000 years or so keeps Chicago archaeologist Richard G. Klein busy counting animal fossils from southern African caves. Those bones tell him as much about the Middle Stone Age humans who ate the animals as about the beasts themselves.

Consider the cave at Klasies River mouth on the Indian Ocean, some 120 kilometers west of Port Elizabeth. Its older levels, dating from roughly 130,000 years ago, contain numerous seashells and bones, including those of fur seals and jackass penguins. This unique deposit, Klein points out, is "the oldest known evidence in the world for the systematic use of aquatic resources by people."

But the deposit contains very few bones of fish and flying birds like gulls and gannets, which, Klein observes, must surely have been abundant near the cave. Bones of fish and flying birds appear only in comparable coastal cave deposits dating from tens of thousands of years later, suggesting that it was not until relatively recently that people developed the technology for active fishing and fowling. The same coastal cave deposits which contain abundant remains of fish and flying birds also provide relics of this technology: stone-line or net sinkers and toothpick-like bone "gorges" which could have been baited and attached to a line for catching fish or birds.

Bones of land animals, too, reflect

man's evolving cultural sophistication. Down to 40,000 years ago, the commonest large land animal at Klein's sites is the docile, cow-like eland. Thereafter, eland bones become much less numerous and those of ferocious wild pigs increase. The implication is that people had developed technological means for capturing dangerous prey at reduced risk to themselves.

"But there's more to it than that," pursues Klein. Besides identifying the bones to species and counting them, he also establishes the ages at which animals were killed. In all his Stone Age sites, he finds, buffalo are represented overwhelmingly by very young and very old individuals, while eland are well represented by prime-of-life adults.

Klein believes the difference reflects the fact that the docile eland are relatively easy to drive; whole herds could be killed in falls over cliffs or in other traps. Buffalo are much more difficult to drive, and it is probably no coincidence that the calves and old individuals that are best represented in the Stone Age sites are the same age classes that are known to be most vulnerable to lion predation. Prime-of-life buffalo are virtually immune to lion predation because of their large size, ferocity and membership in large herds. The same features apparently restricted predation by Stone Age people.

Klein's style of fossil-hunting—not collecting choice specimens, but statistically sampling—is the antithesis of what Karl Butzer decries as "fossilmanship." In the "naughty age" of 19th-century archaeology, Butzer recalls, amateur and professional diggers alike "were seeking pots of gold and museum-worthy mummies." Even today, he remarks, too much research priority is accorded "being first to brandish the 'right' fossil skull or limb bone."

This still-rampant "bone fever" is giving way to recognition by more and more workers in the field and the lab, concludes Butzer, that "anthropology is the science of *context*—a whole vista of trying to see how ancient man lived in a *site* of 200 square feet, in an *environment* of two square kilometers, plus the vegetation and resources he saw in a larger *habitat*—say the size of an Illinois prairie."

In teasing out the roots of humankind's family tree, paleoanthropology is no longer overlooking the forest for the tree.●

National Science Foundation support of research reported in this article is through its Anthropology and Ecology Programs.

Form and Function; The Anatomists View

The shape of a bone, the location of a muscle, the wear on a tooth can help define environmental context and the course of species evolution.

Since the beginnings of paleontology, scientists trying to reconstruct the forms of prehistoric life have had to depend on discoveries of fossil bones and their own best estimates of how those bones probably fit together. Seldom are all the bones found for any one creature. It is as if one were trying to bring out the picture hidden in a join-the-numbered-dots puzzle, with most of the numbers missing.

Additionally, bones alone do not define a vertebrate; appearance and lifestyle are both likely to relate as much to the soft tissue of muscle planes, cartilage and distribution of body mass as to the frame on which those tissues are hung. Such data are seldom available. Only in the discoveries of frozen mammoths have paleontologists so fully fleshed a picture of a prehistoric life form.

Paleoanthropologists readily concede that their anatomical conclusions are often beset both by the subjectivity of the scientist and by ambiguity of the evidence. "Looking at bones can provide as many views as there are viewers," remarks Alan Mann, associate professor of anthropology at the University of Pennsylvania.

Teeth, for example, are a crucial part of the distinction between hominids and pongids. The first lower premolar, for instance, is bicuspid—two-pointed—in hominids and single-pointed in pongids. (*Australopithecus afarensis*, however, regarded as hominid for its bipedalism and other reasons [see "Biology, Behavior and Environment in Human Evolution," in this *Mosaic*] does have a single-pointed premolar in some cases.) But Mann recalls that, as recently as the 1973 International Congress of Anthropological and Ethnological Sciences in Chicago, there was considerable disagreement on the description of a ramapithecine first lower premolar.

In papers from the conference, Mann found descriptions of that tooth ranging

Computer-derived bone. A ball and wire model that represents the computer analysis of bone shape, using multivariate analysis, in the matching of form to function in fossil interpretations.

Charles Oxnard

13

from "semi-sectorial and bicuspid" to "unicuspidate." One author even denied that any first lower premolar of *Ramapithecus* had been found. "There were many different interpretations of fossils that were classified as *Ramapithecus*," Mann recalls.

Paleoanthropologists today, however, are not so limited in their efforts to flesh out the fossil skeletons as were even their fairly recent predecessors. Techniques and perspectives offered by fields as diverse as electron microscopy, biomechanics and paleoecology, now being brought to bear on the question in conjunction with each other, are reshaping not only the conclusions of paleoanthropology but the way the quest itself is being carried forward. Emerging insights derived from such evidence as the microscopic pattern of wear on a tooth, for example, or the kind of muscular stresses that contribute to the form a limb might take, are permitting investigators to derive a new kind of inference from fossil evidence.

Emerging is a pattern of postulates that marry form to function and tell more not only about how a hominid or pre-hominid might have looked, but more as well about how one might have lived—in what environment and under what evolution-directing stresses.

What teeth can tell

The strong differences in durability among skeletal parts means that taxonomic classification of once-whole individuals often is made on the smallest proportion of their parts. Teeth are plentiful; the enamel is a great protector. Mandibles also endure well. But smaller bones of the hands and feet, lower arms, spine and the large-appearing but porous and frangible pelvis do not often appear in the hardened silt matrices of paleoanthropologic discoveries. Even skulls, which we tend to associate most strongly with these discoveries, seldom are found intact and often are missing the bones that would furnish a foundation for extrapolation to facial appearance.

The piecemeal nature of paleoanthropologic material thus precludes a great many possibilities for linking definitely one or another early hominid line to *Homo erectus*, which flourished a million years ago. Before him came several australopithecine species, two of which coexisted about 1.5 million years ago. To judge by the size of their skulls—not brain volume but skull, including all the appurtenances for chewing muscles—one is generally said to have been small,

now termed "gracile," and the other relatively large, "robust."

A problem with these two species is that head and body (post-cranial) fossils have not been irrefutably joined. Owen Lovejoy, professor of anthropology at Kent State University in Ohio, describes the situation: "As far as I can see, there's no way of distinguishing different types of australopithecines from their post-cranial skeletons, and we don't have satisfactory cranial and post-cranial fossils for any single *Australopithecus*. A femur alone could belong to a gracile, robust, or even erectus." Femurs—thigh bones—are the biggest and heaviest of the long (arm and leg) bones and thus the most frequently found.

Whatever their size differences, robust australopithecines shared with the smaller, gracile line a decided change in chewing dynamics from the earliest to the latest of their fossil specimens. The generalized australopithecine condition, says Milford Wolpoff, a University of Michigan anthropologist, is that a shift occurred to a diet requiring powerful chewing. As both responded to the challenge, they narrowed the gap between them. "While one maintained a constant cranial size and expanded its masticatory apparatus," says Wolpoff, "the other expanded its cranial size and maintained what was already a robust masticatory apparatus."

He has conducted a number of studies of macroscopic tooth wear in fossil specimens of various australopithecines, which are generally regarded as having lived in a time when heavy, lush forests were receding and rolling plains of drier savanna grasslands were pushing against the forest borders.

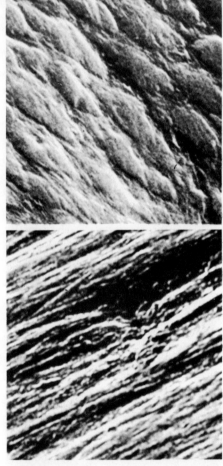

Dietary evidence. Wear patterns on the tooth of a *Hyrax* tell whether it is the tooth of the browsing variety, *H. brucei* (center), or of the grazing *P. johnstoni* (bottom). Similarly, wear patterns identify diet patterns of (from top, center column) a chimpanzee, baboon, *Australopithecus robustus* and *Ramapithecus* and (from top, far right) a grazing warthog, an omnivorous bush pig, a hyena and a cheetah.

Alan Walker, copyright 1978 by the American Association for the Advancement of Science

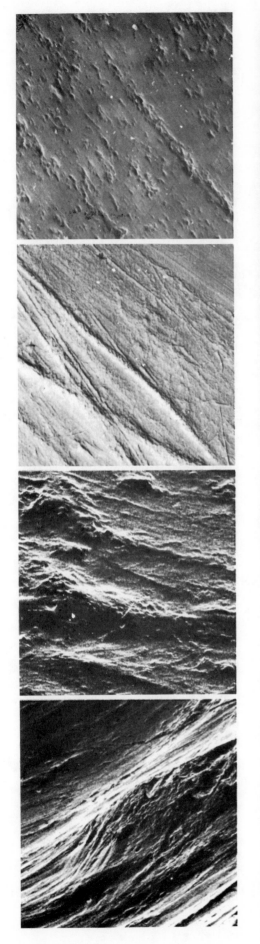

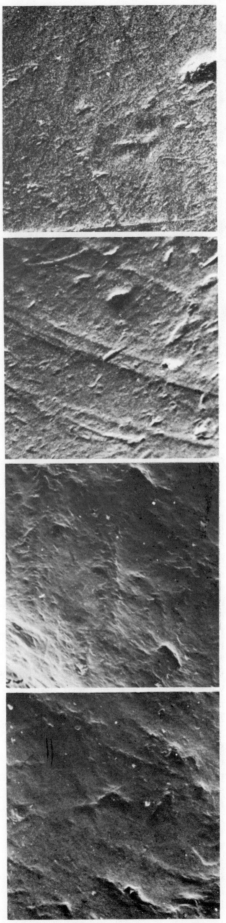

What Wolpoff sees in three age strata of australopithecines is a progression from the terminal stages of a strong cutting function and morphology at least 3.8 million years ago to a situation in which "neither that function nor morphology remains 2.8 million years ago; it's gone in about a million years," replaced by a heavy crunching or grinding function. He regards the change, however, not so much as an adaptation to a diet of seedy fruit as a development that broadened the range of diet to include, for instance, roots on the savanna during the dry season.

More detailed dietary information may be available on the microscopic scale, at which, among others, Al Ryan at the University of Michigan and Alan Walker, professor of cell biology and anatomy at the Johns Hopkins University School of Medicine in Baltimore, have been working. Walker has been pursuing the likelihood that prehistoric teeth would retain microscopic evidence of what its owner ate.

Two kinds of coney

Direct comparisons of fossil teeth with teeth of present-day mammals, whose diets are known, raises many opportunities for ambiguous interpretation. So Walker and his associates began by seeking among living species two that would have thoroughly known feeding habits, be from wild populations, be of similar size, have similar teeth and masticatory systems and live in the same local area so as to minimize the impact of climate and soil differences on their diets. The first step was simply to determine whether animals that matched those criteria would show clear differences in tooth appearance under the great magnification of the scanning electron microscope. Then, if differences were found, they could try to make comparisons with fossil teeth.

After some difficulty in finding mammals that matched all of the criteria, they came up with two species of hyrax or coney, a distant, rabbit-sized relative of the elephant and sea cow. The two species often live in the same burrows but have different feeding preferences. One is strongly a browser, eating vegetation from bushes and trees and very seldom eating grass. The other is mainly a grazer, eating grass when it is available and switching to browsing only when the dry season reduces the succulence and supply of grass.

In the scanning electron microscope, teeth from the two species were clearly different in their patterns of wear, Walker

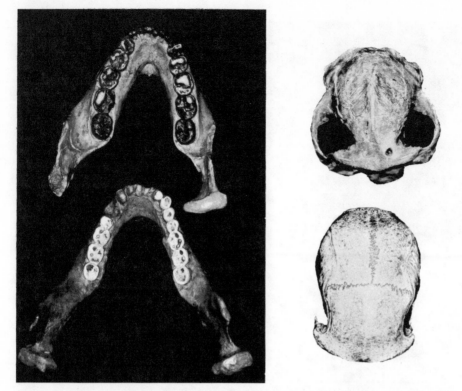

Jaw power. Mandible of a hyperrobust *Australopithecus* (top left) and of *Homo erectus* (below) contrast development and strength of grinding equipment. Hyperrobust skull (top right) and early *Homo sapiens* skull contrast allowance for grinding muscles.

Milford Wolpoff

found. The browser had quite smooth enamel and dentine surfaces, as if the contacting faces and edges had been finely polished. The harder, more organized material of the tooth stood out plainly. Tooth enamel is made up of calcium phosphate prisms, and the dentine within or under the enamel has tiny tubules; the polishing, presumed to be accomplished by cellulose and lignin of the browse, brought this microstructure into sharp relief.

But the species that grazed in the wet season and switched to browsing only in the dry had tooth appearances that varied with the season. After a period of grazing, the teeth would show heavy wear, largely of parallel microscratches that obscured enamel prism outlines and dentine tubular structures. Teeth examined after a switch to browsing would appear more like those of the browsing species, though some evidence of the scratches remained even after the dietary polishing.

The heavy wear and scratching of the grazers' teeth is believed by Walker to result from abrasion by opaline phytoliths (plant opal), silica-laden regimes found most commonly in grasses. Examination of fecal pellets from the two hyrax species disclosed an abundance of phytoliths only in the grazers. Pellets from both species contained tiny rock-dust particles that probably were splashed on grass and bushes alike during rainstorms.

At the very least, Walker says, the differences found between the two hyrax species should provide a basis for information about the diet of a prehistoric individual whose fossilized teeth are in reasonable condition. Walker and his colleagues, before they found the hyrax species, had examined teeth from a wide assortment of meat-eaters, bone-crunchers and omnivores, and have evidence for some tooth micro-appearances that go with the various diets. They now have studied fossil teeth from nearly a dozen robust australopithecines and tend to think that those individuals, which were plentiful a little more than 1.5 million years ago, were probably largely fruit eaters.

"That is counter-intuitive," Walker says, "because they have great big grinding surfaces on their molars, and the presumption would be of very forceful grinding action. But a big grinding area requires big masticatory muscles to apply great pressure. And when you estimate the masticatory muscle mass from the skull appearance, the grinding pressure would not seem to be unusually great. We gather that the big grinding area indicates the processing of a large amount of food—little, seedy fruits without the meaty mass we associate with fruits today and of poor nutritive value compared with today's."

The hominid fossil work is only beginning, and Walker wants to increase greatly the number of samples in order to reduce ambiguity. He is working toward a process—mainly dependent on a new generation of scanning electron microscope—that will be automated, provide high-contrast images for photoanalysis by machine and enable a relatively bias-free comparison of specimens by investigators at different laboratories.

Hip and thigh

Fossil teeth may be more numerous than any other collected remains of prehistoric hominids, but the greatest intensity of function study probably has been not on chewing but on locomotion. Most of us see as a major difference between apes and humans the fact that humans—or protohumans—came down out of the trees and walked upright. Paleoanthropologists have much the same view, but—and the sequence is important—they propose that we were well into being hominid when we became "post-arboreal" and then unremittingly "bipedal."

Evidence that might tell when and in what order these humanizing events occurred has been studied in several ways, but none has provided a completely unarguable conclusion. One technique is morphometrics, which begins with a series of detailed measurements of a fossil bone in a variety of aspects. The measurements then are subjected to analysis by means of computer techniques, such as linear regression and multivariate statistical analysis.

Morphometric analysis to deduce locomotion patterns is most associated with Charles Oxnard, dean of the graduate school and professor of anatomy and biology at the University of Southern California. He has used the analyses to look for similarities of hominid fossils to living nonhuman species in order to infer the prehistoric function from the nearest present-day analogue. These and other modern analytic and experimental procedures, including engineering stress and laser optical data analyses, have led Oxnard to suggest that australopithecines, in their locomotor parts, had morphologic resemblance more to modern orangutans than to other apes, which would imply abilities for climbing trees. This is not to imply that there is any genetic relationship between *Australopithecus* and the orangutan, Oxnard notes. It is simply that the ancient

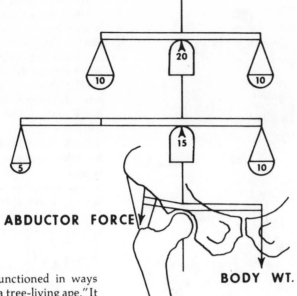

ABDUCTOR FORCE

BODY WT.

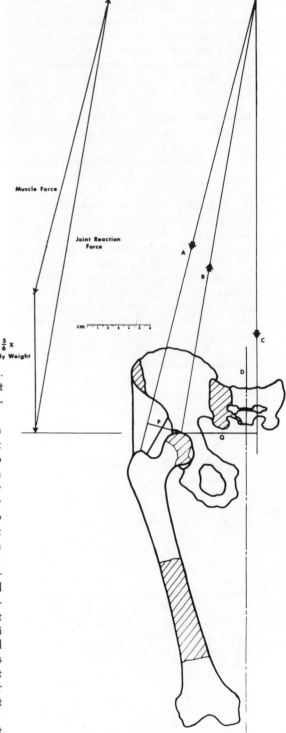

hominid may have functioned in ways "that are paralleled in a tree-living ape." It is Oxnard's view that the biomechanics of australopithecine bipedality were different—and therefore differently derived—than in modern man.

Owen Lovejoy at Kent State University also draws on biomechanics in exploring the question of australopithecine locomotion, but arrives at a different conclusion: He tends to connect australopithecine and modern two-leggedness. Lovejoy and his co-workers published five years ago their conclusion that *Australopithecus*, from the earliest example, was completely bipedal. "I think that at least since 3.5 million years ago all hominids have been bipedal," he says.

The calculations that go into that conclusion include a balance-beam diagram and equation that characterize the distance between hip sockets and the length of the necks of thigh bones, the ends of which fit into those pelvic sockets. Compared with modern humans, *Australopithecus* had long-necked femurs. But he, or more appropriately she, also had a narrow pelvis and a narrow birth canal for small-headed progeny. Lovejoy believes this combination gave the australopithecines of two million years ago an adaptation to bipedal gait superior to ours because the structure demanded less force from abductor muscles to keep the pelvis level when one foot is raised off the ground.

Adept bipedality, in that context, begins to seem less closely associated with the drift toward erect humans, fleet of foot and with hands free to carry weapons of the hunt. "Bipedality came well before hunting," Lovejoy says. "Australopithecines probably did some scavenging of carnivore left-overs, but you have

to remember: They really were animals. Not just less intelligent versions of us, but hardly more intelligent than chimpanzees."

But even if those australopithecines were bipedal, were all subsequent australopithecines bipedal? And was so human an attribute to descend to *Homo* in a line so straight as to enable us to identify any creature as truly ancestral to any other? The questions would seem to underscore the analogy of the dot-to-dot puzzle drawing, most of the dots of which are missing.

Charles Oxnard, for instance, compares the evidence of 4-million-year-old australopithecine bipedality with evidence drawn from articulation of a foot dated at 1.7 million years from Olduvai Gorge and assigned to *Australopithecus* "and indeed assigned even closer to man, as *Homo habilis*, by some workers." In that foot he finds reason to question later hominid bipedality, at least in that specimen.

The remains of the foot—the only foot that has so far been studied—says Oxnard, "have been articulated so that they seem to possess the high transverse arches that are distinctive of the human foot." But his own recent work reveals that rearticulation "is incorrect."

"When aligned correctly," Oxnard says, "this foot does not resemble that of man; though the creature may have walked upright, there is far more anatomical evidence that the foot was used in a climbing mode. Thus, although footprints

Hip and thigh. Critical dimensions of the apparatus (top left) used to determine control of pelvic tilt and efficiency of gait in *Australopithecus*. Frontal projection (above) and force diagram for an australopithecine pelvis and femur show hip joint pressures to be only half those of humans.

C. Owen Lovejoy, "The Gait of *Australopithecus*." Reproduced by permission of the American Association of Physical Anthropologists from the *Yearbook of Physical Anthropologists*, 17:157 and 158, 1973.

at nearly 4 million years may indicate bipedality, the particular foot of the australopithecine from Olduvai, at 1.7 million years, is something else."

The line (or lines) between 4 million and 1.7 million, then, is (or are) to be defined by still-missing dots.

Muscle structure

Other scientists, employing modern anatomical approaches to the investigation of locomotion, put muscles on those bones in the effort to shed light on whether a particular hominid was built for bipedality. Such is the work of functional morphologist Jack Stern, associate professor in the Department of Anatomical Sciences at the State University of New York, Stony Brook. One of his studies sought to discover why human beings are equipped with a buttock muscle, the gluteus maximus, that differs from that of non-human primates by having a bigger upper part, arising from the hip bone and having its insertion in the side of the thigh.

"For both humans and other primates," Stern says, "the lower part of the gluteus maximus sweeps the leg back, which is important for fast-dash running. But it is a motion that doesn't require the upper part of the muscle. That upper part is clearly an abductor muscle, secondary to the gluteus medius, helping to keep the pelvis level when one foot is off the ground. It turns out that the upper gluteus maximus has a special kind of abductor role," Stern says, "to assure a reasonable pelvic levelness when the raised foot comes down in slow running or jogging." He can envision that muscle as making the difference between success or failure in the type of hunting that Australian aborigines still do: running a prey to ground, either by injuring it first or by keeping it moving so it can't stop for food.

The way that Stern and his co-workers find out which muscles are doing what is to take electromyographic measurements of muscles during movement. Small electrodes in the muscle detect any difference in electrical potential when the muscle contracts. At Stony Brook, the investigators record electromyographic information by telemetry from free-moving animals carrying small transmitters in back packs.

In this fashion, it has been possible to study the contractions of specialized muscles that some non-human primates share with humans and to determine when they use them. In the main, Stern says, the muscles humans have in com-

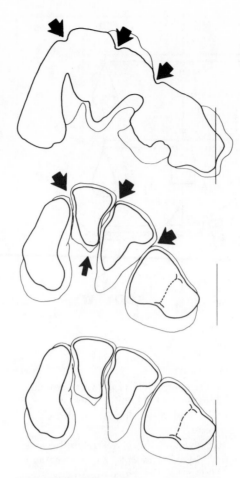

Rearticulation. Cross section of an Olduvai foot shows irregularities if articulated to form a human-like arch. Later "stepless" profile has lower arch.

Charles Oxnard

mon with other primates are employed most strenuously in slow climbing and not in walking on all fours or swinging by the arms. Neither do humans have much kinship of muscle contraction with the running and jumping activities characteristic of species such as squirrel monkeys. Stern and John Fleagle, also an associate professor at Stony Brook, have attached strain gauges to the collar bones of non-human primates to determine why clavicles are shaped as they are and what their function is. One answer is that clavicles of primates serve to keep the shoulders from collapsing when the animal walks quadripedally on a tree branch.

Time to learn

When it comes to studying anatomical function in early hominids, some importance attaches to their stages of aging, their physiologic development with age and their individual lifespans. As Alan Mann points out, "The fossils represent the way they were when they died—at whatever age."

Mann went at age-staging by comparing times of molar eruption between today's human being and today's chimpanzee, and then looking at *Australopithecus*. Chimps get their first molars at about three years, he says, while we get ours at about six. That roughly twofold time difference between chimps and human beings holds for several sets of tooth eruptions.

X-ray studies of fossil australopithecine jaws, the state of dental eruption in which suggests an age similar to that of today's first-grader, showed the same spacing of molar eruption as that of a modern child. The fossils have unerupted teeth in place and, Mann says, an erupted first molar accompanied by only the smallest bud of an unerupted second molar. There is no sign of a third molar. Chimpanzees, with shorter spacing of molar eruption, will have all molars visible to X-ray after the first one has erupted.

The delay in the australopithecine tooth eruption schedule, compared to that of chimpanzees, means to Mann that the hominid had a prolonged childhood, even though australopithecines and chimps have a comparable brain size.

What is the value of a prolonged childhood? "Time to learn," Mann suggests. "Behavior of australopithecines was complex enough even then, I believe, that it took time to transmit it from parent to offspring. It took an extended dependency."

At Michigan, Milford Wolpoff says the studies by Mann and others indicating delayed maturation of australopithecines also suggest they "had menarche not much sooner than we do," which would mean a reproductive age beginning at about 12 years. But the hominid "lived beyond sexual maturity only 13 years or so as an adult," he says, "so that if an australopithecine couple had its first-born when the mother was 13 years old, there was a 50-50 chance that both parents would be dead before they could finish rearing a third offspring—before they could do any more than reproduce their own number and have it learn anything."

But those hominids obviously did manage to build a population that grew in number beyond a few breeding pairs. What, then, was the learning resource for orphaned offspring? Wolpoff, in what he declares is an "off the wall theory," believes that australopithecines "had to recognize lateral kinship—aunts and uncles—as adults from whom to learn." He points out that maturation delay was increased further from genus *Homo*, as expressed in the Neanderthal of 100,000

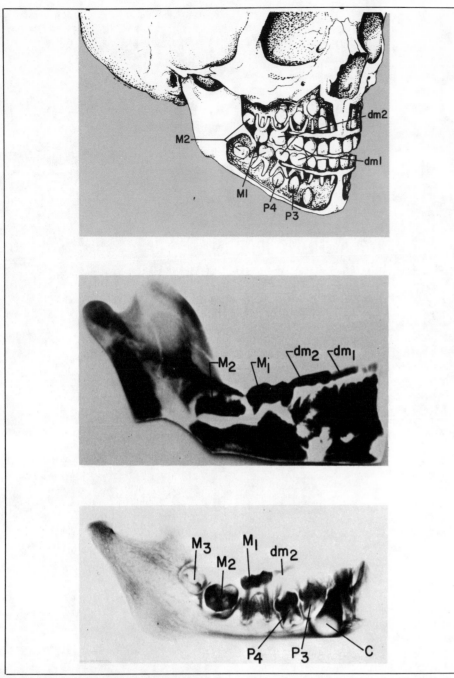

Molar eruption. Appearance of permanent molars in a modern child (top) is more like that of a young australopithecine from Swartkrans (center) than that of a faster developing young chimpanzee (bottom) is like either.

Alan Mann

years ago, to modern *Homo sapiens*. Neanderthal got its third molars at about age 15; we get ours, if we get them at all, at age 18 to 20. Wolpoff thinks the third-molar lag in us as compared to Neanderthals probably occurred in a relative "jump," and is looking to find when it occurred.

Pair bonding

Owen Lovejoy, with the same maturation and reproduction figures in mind for *Australopithecus*, tends to emphasize that the hominids could have had multiple dependent offspring—several juveniles in tow and learning at the same time. Human beings are unique among the primates in having no estrous cycle, he notes. Although that situation might contribute to a possibility of having more offspring during a short life, it only becomes an advantage if mating individuals form a special type of stable pair bond, an arrangement in which a male is provisioning a female to help assure that his offspring will survive, while she finds that her best chance, and her offspring's, is to stay with the good provider.

If one looks at it that way, says Lovejoy, then the strong sexuality of human beings is best expressed "not in attracting the opposite sex but in maintaining a pair bond."

To carry it to the logical cultural conclusion, Lovejoy says, such pair-bonding would constitute a primitive family, which contributes to a social structure, which improves the chances for survival as compared with the condition of individuals and mother-offspring pairs in a nonfamily structure. "That's how we got to be long-lived, because the social structure developed along with reproduction of multiple dependent offspring," he suggests.

The conjectures of Wolpoff and Lovejoy about the behavior patterns that might have accompanied the deduced functions of prehistoric hominids apparently are common among paleoanthropologists. Alan Mann is working on a book in which he will describe the contribution of studies of human and other primate evolution to an understanding of current human social problems.

"I don't think analysis of form to function is enough," he says. "You have to carry the function to successful behavior. We know that our ancestors, however far back, were successful, because we're here. But how did a particular functional complex contribute to a behavior that was successful? As a very basic for-instance, why did bipedalism evolve?

"We learn from the people around us. Millions of years of successful primate evolution provide the basis for learning a core of stereotyped reactions to rely on in relating to other individuals in society. And it worked well through the hunter-gatherer phase leading to modern *Homo sapiens*. But with the coming of agriculture, the delayed maturation period had us learning from the wrong examples, sometimes inappropriate behavior for the context.

"What doesn't work today," Mann says, "is the ability of individual human beings to predict the actions of others. The yawn of a baboon means something to another baboon, and elicits an appropriate response from it. But a woman's smile to a man in our society can mean any of dozens of things."•

National Science Foundation support of research discussed in this article has been through its Anthropology and Systematic Biology Programs.

Shifting Perspectives on Early Man

Dental adaptation to diet shifts caused by climatic change link *Ramapithecus* to *Homo sapien,* over a span of a dozen million years.

For modern humankind, so aware of its own position in time as well as in space, the lure of the past is of more than casual concern. The need to know where we came from can be at least as strong as the need to know where we are going. In the process of learning one, in fact, we may gain insight to the other.

Current research is finding human origins farther and farther back in time, some dozen million years now beyond what had been thought. The scepter of primacy is currently being carried by a diminutive hominid found in several forms on several continents.

This antiquity of human origins, however, may not be the most startling or even the most significant element produced by the new perspectives modern studies are bringing to the understanding of human origins. More important may well be the increasing body of evidence to the effect that the initial revolution that began the line to man was exceedingly humble; that in the processes of change, as well as in the nature of changes themselves, the creatures that led to man were more like than unlike many other creatures, two-legged and four, with which they shared their living space.

We are coming to know our forebears from a mass of evidence accumulated during the past 50 years or so, with the decade just past being an especially rich time. Through this most recent period, fossil remains and entire, new, fossil-rich areas have been found not only in Africa, but in Pakistan, China, Java and other parts of the Old World as well. Uncovered have been an estimated 1,800 specimens, including everything from isolated teeth and skeletal fragments to complete or nearly complete mandibles, femurs and skulls representing perhaps 500 or 600 individuals who died at widely scattered times and places.

Interpreting these remains is an ongoing process. Each new find is not only important in itself, but also affects the way scientists analyze and reanalyze former finds. Since man was shaped by his environment and his relationship with other species, geological formations are being searched not only for remains and tools of early man, but for clues as well to the climates and landscapes in which he lived and fossil remains of species that hunted and were hunted by him. The object is to reconstruct not only the appearance of our remote ancestors and their home ranges, but also their behavior and, beyond that if possible, their life styles.

"In History, a past bespeaks a future; a future imperatively demands a past." George Dangerfield in *The Damnable Question*.

The uncovered past. Glynn Ll. Isaac (left) uncovers antelope pelvic bone east of Lake Rudolf, strengthening the idea that Pleistocene hominids were carrying food back to a home base. A young African boy (above) uses a stone flake to slit the skin of an antelope carcass.

A humble beginning

The beginning was humble indeed. Early signs of the coming of man, ultimately the thinker supreme among living things, seem to have been as unprepossessing as changes in tooth structure. Basic changes in the brain came much later.

The creature now often considered to be the original "hominid," the first member of the family of man, was discovered in northeastern India in 1932 and is known as *Ramapithecus.* (*Rama* is a hero in Hindu mythology, and *"pithecus"* is Greek for "ape.") A small-brained creature, it might have stood about three feet tall and weighed about 40 pounds or so.

As recently as 1961, *Webster's Third New International Dictionary* was calling *Ramapithecus* "a genus of Upper Pliocene Indian apes...exhibiting almost human dentition." The Pliocene period ended some two million years ago. *Ramapithecus* is now known to be both much older and more widespread, as well as, on the basis of current thinking, far more human.

As much as ten million years or more were added to its age when, in the early sixties, the late Louis Leakey, director of the National Museum of Kenya, excavated part of an upper ramapithecine jawbone and several teeth at a site near Fort Ternan east of Lake Victoria. Geologists at the University of California at Berkeley dated the fossils by dating samples of volcanic rock in the same deposits. Such rocks contain atoms of radioactive potassium which break up spontaneously and produce the inert gas argon. Trapped in rock crystals after the lava hardens, the gas accumulates at a steady rate and can be measured by devices sensitive enough to detect a billionth of an ounce of material. This method indicates that the Fort Ternan fossils take *Ramapithecus* back 14 million years, give or take a few hundred millennia or so.

Short in the tooth

So far, our only traces of *Ramapithecus* appear to be facial remains: more than 50 jaw fragments and teeth, of which molars and premolars provide clues of special interest. These are massive teeth, probably up to 20 percent bigger than the back teeth of most similarly sized apes, with crowns covered by an extra-thick layer of enamel. The lower jaw is also thickened and buttressed. In addition, the back teeth may have one or two added points, or cusps; the premolars tend to be "molarized," with flatter and more-rounded cusps. The canine teeth are short and small, rather than long and tusklike as in many primates; the incisors are tiny. And what ideas of any real significance can such an array of dental characteristics point to? Properly read, quite a lot: According to Clifford Jolly of New York University, one change of which teeth can tell is dietary adaptation, a basic shift in one's eating habits, which means a change in the environment and considerably more. "The dental structure of most of the apes that were living in *Ramapithecus'* times," Jolly explains, "was balanced, with back teeth large enough to deal with leafy vegetation and tough stems, and incisors for slicing and peeling fruits. The teeth of *Ramapithecus*, however, add up to more powerful chewing with the back teeth and less food preparation with the incisors." Since long upper and lower canines tend to interlock, these teeth were reduced in most ramapithecines, thus permitting more effective side-to-side grinding jaw movements. (Hungarian ramapithecines appeared to manage without shortened canines, Milford Wolpoff at the University of Michigan points out; also, notes David Pilbeam of Yale University, some apes manage side-to-side grinding with large canines.)

Nevertheless, only by examining these changes within the context of other things that were happening in the world at the time, can their real significance for the history of humankind be made to emerge.

There were, for instance, some major geological changes taking place at the time. After ages of warm climates, world temperatures had begun to tumble. One notable result was a dwindling of the dense, moist, evergreen forests that covered most of the continents, housing a rich assortment of leaf- and fruit-eating primates. In their place came the fruits of drier times and climes: a spread of open woodlands and grassy savannas. Plants typical of such terrain include tough roots, tubers, hard seeds and nuts, all of which called for precisely the sort of heavy-duty, crushing-grinding teeth found in *Ramapithecus*.

Apparently, Jolly and his colleagues point out, in the face of broad environmental change, *Ramapithecus* had abandoned the dwindling forest resource to his pongid contemporaries, and stuck out to build a new life in the encroaching grasslands. The accompanying dietary (and dental) adaptations to savanna life apparently gave *Ramapithecus* the tie he needed to his long and human future.

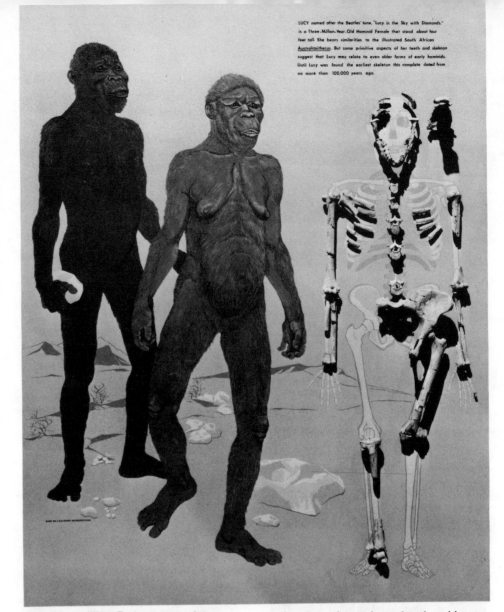

LUCY named after the Beatles' tune, "Lucy in the Sky with Diamonds," is a Three-Million-Year-Old Hominid Female that stood about four feet tall. She bears similarities to the illustrated South African *Australopithecus*. But some primitive aspects of her teeth and skeleton suggest that Lucy may relate to even older forms of early hominids. Until Lucy was found the earliest skeleton this complete dated from no more than 100,000 years ago.

Age: three million. Forty percent of the skeleton of australopithecine female and what she and her mate might have looked like, as reconstructed in a display case at the Cleveland Museum of Natural History.

Ramapithecus was surely not the only creature that had to adapt. Alan Walker of Harvard University points out that at Fort Ternan, for example, other species, presumably the descendants of forest-dwelling forms, were also in the process of entering open country and exploiting savanna-type foods. Among them were the ancestors of modern ostriches and aardvarks, small antelope with spiral horns, a slender and short-limbed giraffe-in-the-making and a hyena as large as a grizzly bear and equipped with canines up to eight inches long in a three-and-a-half-foot skull.

The movement was apparently worldwide. Open-country species appeared wherever forests were on the decline. Specimens of *Ramapithecus* have been unearthed in Greece, Turkey and Hungary, as well as in East Africa and India. But the richest fossil yields are coming from the Siwalik Hills of northern Pakistan, in the same general region where the original specimens were found more than four decades ago. In three seasons of full-scale fossil hunting there, Yale's Pilbeam and his associates—an international team of investigators from France, England, the United States and Pakistan—have found 86 new primate specimens (out of a total of more than 13,000 fossils of other animals), doubling the number previously found in the region.

The new evidence has brought a more sophisticated view of hominid evolution, a greater sense of the complex of forces which gave rise to our kind. Pilbeam stresses that an as yet undetermined number of such species, all representing *Ramapithecus* and related primates, existed: enough to make up an entire new category of primate—a separate family of pre-man or incipient hominids—to go along with the many previously recognized pongid and hominid families. According to Pilbeam and his former student, Glenn Conroy, now of New York University, *Ramapithecus* is the form "most plausibly interpreted" as the first stage in human evolution.

But the Yale investigator is concerned with far more than primate taxonomy, the art of classifying primates: "I used to look at early hominids as if they were links, important simply as possible steps on the way to man. Now I also appreciate them for their own sake, and see how they interacted with other primates and other mammals. Now I'm less interested in family trees than in behavior and the dynamics of adaptation."

The next step

An enormous time gap exists between today's leading candidate for hominid number one and subsequent stages in human evolution. *Ramapithecus* became extinct about eight million years ago, and reasonably abundant traces of hominid number two do not show up in the fossil record for another four to five million years. But a great deal happened in the interim. It was partly a matter of continuing ramapithecine trends, adapting more completely to life on the savanna by developing more powerful chewing mechanisms and exploiting a greater variety of foods. Also, there was a premium on increasing size. Hominids larger than average enjoyed advantages over their less favored brethren when it came to discouraging attacks from big cats and from other predators. From the standpoint of the future, however, the most significant changes were changes in the structure of the brain.

Hominid number two was discovered more than half a century ago by Raymond Dart of the University of the Witwatersrand in Johannesburg, South Africa. Dart called it *"Australopithecus,"* or "southern ape." But he knew it was more than an ape. He identified it as "an ultrasimian and prehuman stock," though the verdict was not generally accepted for another generation. Since then, further specimens of australopithecine and more advanced early hominids have come from other South African sites and, more recently, from sites in East Africa.

In 1959, large-scale digging began in the Olduvai Gorge of Tanzania, when Mary Leakey, wife of Louis and an experienced investigator in her own right, found parts of a massive skull exposed in the side of a cliff. Other sites yielding important remains include the Omo River Valley in southern Ethiopia, explored by Clark Howell of the University of California at

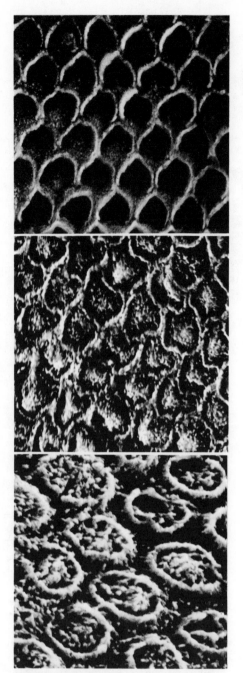

Humble beginnings. Scanning electron micrographs of tooth enamel prism patterns of (from top) *Ramapithecus*, *Homo sapien* and the ape *Pongo pygmaeus* show effect of shifting diet forced by shift in environment and living patterns. Those of *Ramapithecus* are more like man's than like ape's. (magnifications: about 2,000; enlargement factor, 1:8)

Berkeley and his associates, and a region in northern Ethiopia, not far from the Red Sea, which is the fossil-hunting grounds of Donald Johanson of the Cleveland Museum of Natural History and Maurice Taieb of the National Center of Scientific Research in Paris. Another member of the Leakey family, Richard, who has taken his father's place as director of the National Museum of Kenya, works with an inter-

national team in the East Rudolf desert region near Lake Turkana (formerly Lake Rudolf).

Evidence gathered to date from all these sites includes skulls, plus jaws and teeth and much of the rest of the skeleton. The remains represent at least 400 individuals and several different species. The earliest known australopithecine individuals averaged about two feet taller than *Ramapithecus* and were perhaps three times as heavy. Like the earlier primate, they seem to have been living on tough, open-country foods. Their back teeth were also large, featuring molars about twice the size of even ramapithecine molars (and double the size of ours) and large molarized or flattened premolars. Tim White of the University of Michigan has studied the muscles that moved these grinding-crushing elements. Forces acting on the back teeth required special bony buttresses, to counteract compression of the lower jaw, and extraheavy flaring cheek bones for the attachment of large chewing muscles. The new effect was a deep-set, wide face.

The hands of these strange-looking creatures seem to have been much like ours, and they may actually have walked somewhat better than we do. According to Owen Lovejoy of Kent State University, the *Australopithecus* pelvic region was better balanced; they would have swayed less and used less energy in striding. They were certainly strong: Wolpoff describes large ridges at the elbow end of the humerus that attest to the attachment of massive muscles. The thighbone, a thick, dense shaft with a small, marrow-containing channel running down the center, was an order of magnitude sturdier than ours, which is mostly central channel. Further information about body structure can be expected when Johanson and his colleagues finish analyzing one of their most remarkable finds, a set of associated bones making up 40 percent of the skeleton of a three-million-year-old female *Australopithecus* known as Lucy.

Finally the brain

There was more to these hominids, however, than strength and physical bulk. In them we see the beginnings of the distinguishing mark of the human species, hints of fundamental changes in the species' cerebral organization. The australopithecine brain occupied some 450 to 500 cubic centimeters, double the estimated cranial capacity of *Ramapithecus* and appreciably less than the modern average of some 1,400 cubic centimeters.

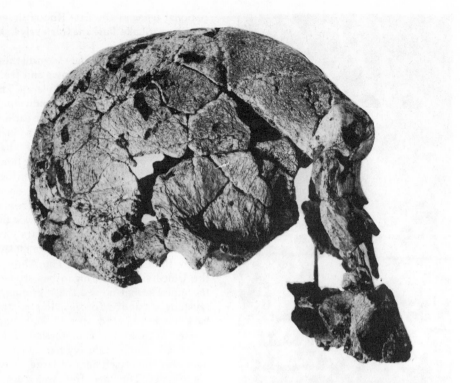

Undisputedly *Homo.* Best profile of painstakingly reconstructed skull ER1470 shows care with which pieces are fitted. Twice as many pieces still await fitting.

But size is by no means the whole story. Subtler insights are being obtained in the laboratory at Columbia University where Ralph Holloway has developed an ingenious procedure for learning more about the structure of prehistoric brains (see "Our Cerebral Heritage," *Mosaic*, Volume 7, Number 2).

Starting out with an intact or a reconstructed skull, he holds it upside down and pours liquid rubber latex into a hole at the bottom, just enough to build up a clinging, thin-walled sheet whose contours follow closely the contours of the skull's inner surface. He thus captures the tiny bulges and hollows, the imprints of the brain that once filled the skull. Baking the specimen produces a tough latex lining, a hollow mold which is later filled with plaster and cut away, leaving a solid replica of the brain, remarkable in its surface detail.

Holloway has prepared 140 such "endocasts," representing the brains of monkeys and apes as well as assorted hominids, and 45 of them are being studied in an effort to discover changes in shape that reflect basic evolutionary changes. Using a stereo-plotter, a device adapted by Walker from methods employed by cartographers, to measure cranial dimensions precisely, he plots the information on special charts, feeds it to a computer and ends up with maps indicating the brain-shape contours. In preliminary, though more than cursory study, Holloway has noted an expansion of certain so-called "association areas" on the side and top of the australopithecine brain, areas which perform high-level analyses of signals flowing in from the creature's sense organs and which, some neuroanatomists believe, may be concerned with language. (See "A Window to the Brain,"*Mosaic*, Volume 7, Number 2.)

Such observations do not prove that the hominids were using some sort of language, but Holloway is among those who consider that to be a definite possibility. Also tending to confirm the respectable quality of *Australopithecus's* small brain is the fact that his permanent molar teeth erupted at the same life stages as ours do, the last coming in at about the age of 18. Alan Mann of the University of Pennsylvania, who conducted this study, cites it as an indication of a long period of childhood—time to master "a large complex of learned behavior."

Even in those prehuman days individuals had to acquire an appreciable body of knowledge before taking their places in society. For example, one trend in the evolution of our kind was an increasing emphasis on meat eating. Although primates are predominantly vegetarians, chimpanzees may hunt and, on occasion, hunt cooperatively. But they do not need meat to survive. Things were apparently different on the savanna in remote times, and the remains at early sites of everything from lizards and rats to remains of hare, giant pigs and various pachyderms show that hominids exploited practically every available source of meat.

We do not know as yet how many australopithecine species existed, how many different life styles hominids developed in adapting to ancient wildernesses. We do know, however, that many of the efforts failed. For example, about 2.5 million years ago two candidates for primacy, both "robust" and "hyperrobust" species, appeared in the australopithecine line, creatures that might have been up to double the size of their smaller contemporaries, with molars twice as large and muscles to match. Like present-day gorillas, they may have exploited highland forest zones, using their teeth to shred bamboo grasses and other fibrous plants. Or they might, as some scientists think, have represented some special (and thereby unsuccessful?) savanna adaptation. As far as sheer strength is concerned, they were magnificent primates. But that adaptation led nowhere; they became extinct, for reasons we have not unraveled, about a million years ago.

The take-off period

The future belonged to less-brawny, brainier breeds. As archaeologist Glynn Isaac has reconstructed the two-million-year-old scenario: Imagine a prehistoric gallery forest, a ribbon of green marking the course of a river running over wide, dry savanna grasslands. A few hundred yards away, walking across the open plain toward the trees, are four hunters with sticks in their hands. One of them moves easily under the load of a 150-pound impala slung over his shoulder. As they approach their camp among the trees, a closer view reveals individuals somewhat on the undersized side compared with us, the tallest measuring rather less than five feet. They have large faces with wide, flat cheeks, brows that slope backward and not much of a forehead. Nose, mouth and jaw jut forward to form a distinct muzzle.

Isaac imagines a noisy coming together taking place at the forest's edge. The hunters and the rest of the band, mainly females and their offspring, exchange greetings and proceed to divide the spoils. Everyone is hungry for red meat. All eyes focus on the impala carcass being dismembered with crude knives, razor-sharp flakes struck from convenient stones. Individuals snarl and jostle for position as chunks of meat are passed around, cut into smaller pieces, devoured raw on the spot.

Creatures like these, whose description is a composite of many findings and insights, mark the beginning of the take-off period in man's evolution, the period that saw the most spectacular increases in brain size and social development. They represent the first human beings, the first members of the genus *Homo*—and, according to most investigators, they arose about two million years ago from smaller australopithecine forebears. Perhaps the most striking specimen of the new hominids is a skull catalogued as ER1470 (East Rudolf specimen No. 1470), announced by Richard Leakey in 1972.

The specimen was found in a thoroughly smashed and shattered condition. Reconstructing it amounted to tackling a super jigsaw puzzle in three dimensions. Some of the pieces are missing; others are so small that they had to be fitted into place under a microscope. Nor could the solution be verified since no one has found an intact skull against which the "answer" can be checked. Richard Leakey and his wife, Meave, have spent hours on the puzzle. So has Walker, who is expert at such work. Although several hundred pieces remain unplaced, some 150 have been fitted and glued together. That is enough to reveal an individual who had a large australopithecine face but a cranial capacity, when measured from a Holloway endocast, of 770 to 775 cubic centimeters, half again as big as the average for *Australopithecus*.

Defining man is always a tricky business. But there is general agreement that when populations attain sizes like that of ER1470 they belong in the *Homo* category. Potassium-argon tests indicate that he lived 1.8 million years ago. The almost complete skull of an even bigger-brained individual has recently been discovered, again in East Rudolf, in deposits dated some 200,000 years later. Large enough to house a brain of perhaps 850 cubic centimeters, it is the oldest known specimen of *Homo erectus*, a species which attained a height of five feet plus a few inches, a weight of some 175 pounds and a cranial capacity in the 775- to 1,300-cubic-centimeter range.

Cerebral expansion was one of a number of changes under way during man's take-off period. In a recent study, Wolpoff described other *erectus* features, such as smaller molar and premolar teeth and a sharp increase in front-tooth loading. Signs of the latter development are large overhanging brow ridges—bony buttresses to absorb stresses transmitted upward through jaws and cheeks from incisors and canines—and a muscle-

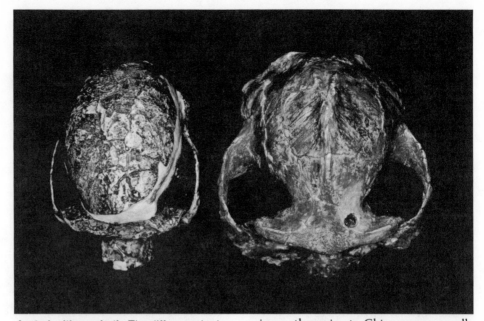

***Australopithecus* both.** The difference in size between hominid species that probably shared locales some 2.5 million years ago can be judged by these casts. The looplike openings at each side of the head, as seen from the top, were the space needed to accommodate the muscles that powered the jaw.

attachment area at the back of the skull two to three times as large as the corresponding area in *Australopithecus*.

Such changes are what one would expect if the front teeth were serving increasingly as tools for gripping, twisting and biting hard. Among other things, they may have been used to hold spear shafts for flexing and straightening or chunks of meat while cutting off bite-size pieces. Another change is a thickening of the skull, which Wolpoff suggests may be "an adaptation to those accidents occurring when one is hunting big animals at close quarters."

The earliest tools

More direct clues to the rise of hunting turn up at the beginning of the take-off period. The world's earliest fabricated stone tools, dating back about two million years, have been found in the Omo River Valley where, according to Howell, "Workshop sites, including choppers, flakes and lots of waste chips, are scattered over an area of several square kilometers." An East Rudolf site excavated by Glynn Isaac of the University of California at Berkeley contains hippopotamus remains, together with several choppers, a hammerstone and other tools presumably used to butcher the animal.

By this time our ancestors had long since developed behavior patterns found in no other primate. Chimpanzees usually sleep where they happen to be when darkness sets in. Baboons may return to the same trees they slept in the night before, but the entire troop always leaves as a unit the next morning. Man alone among primates has a home base, a place where some individuals stay behind to "keep house" and look after the children, while others go out in search of food. Camp sites thick with tools and accumulated bones of small and large game have been excavated at the Olduvai Gorge and elsewhere. They indicate that sharing among closely associated families had become a way of life, a prerequisite for survival.

Toward the end of the take-off period, these people achieved what Isaac calls "a quantum jump in the design of tools." Their ancestors had made tools without any apparent patterns in mind, simply striking a number of flakes off a rock and selecting one which happened to be suitable for the job at hand, say, slicing meat or whittling the business end of a wooden spear. But by 1.5 or so million years ago, some tools were being made rather less casually: They were worked on both sides and along the edges to produce roughly oval or almond-shaped forms.

These so-called "hand axes" signal a new kind of disciplined thinking, the emergence of symmetry as indicated by tools produced to preconceived, standardized patterns. Despite the name, we do not know what they were used for. But they may symbolize the start of a general tightening, formalizing of society. Isaac feels that making tools to a regular design may have been only one of a number of imposed behavior patterns, part of a development which also included food taboos, kinship customs and certain rules of language.

Ages separate the first near-ape hominids from their descendants, the first hand-axe makers. The process that began some 15 million years ago with the spread of open woodlands and savannas and the appearance of *Ramapithecus*, the process that subsequently led four to five million years ago to *Australopithecus* and three to four million years later to *Homo erectus*, brings us most of the way toward modern man. *Homo erectus* was more of a wanderer than was *Australopithecus*. His traces, which include Peking man and Java man as well as various European specimens, predominate among the earliest prehistoric sites of Europe and Asia. Apparently he had developed the technology that is required to cope with climates colder than those prevailing in Africa south of the Sahara—not only the tools and hunting-gathering tactics, but also fire, clothing and shelter, which represent an ability to modify his environment. The earliest known hearths burned in China more than half a million years ago, and they are hearths of *Homo erectus*. The next species in the human sequence is *Homo sapiens* himself, who probably appeared on the scene between 300,000 and 400,000 years ago. That, however, is another story.

A word of caution: The sequence presented here is by no means the last word. It is subject to change without notice, in the light of new research.

Some investigators believe human evolution started later and proceeded faster, or that *Ramapithecus* is not really a hominid, or that hominids originated from an as yet unidentified creature who arose a mere six to seven million rather than 15 million years ago. *Homo* may have appeared much earlier; Johanson has found Ethiopian specimens representing at least ten individuals whom he considers human and who lived three million years ago, more than a million years before ER1470. And Mary Leakey claims human beings at least half a million years older than that from a site near Olduvai.

Efforts are also under way to fill the gap between the vanishing of *Ramapithecus* some eight million years ago and the australopithecine finds of three to four million years ago. Howell, who has just returned from a survey in Libya, where promising fossil deposits exist covering most of that time span, stresses the importance of finding more specimens:

"We've learned an enormous amount since Olduvai was opened up nearly 20 years ago. But human evolution is an unfolding story and we need fresh evidence."

There is also a mounting drive to analyze more intensively what we already have, to extract more information from each specimen and from large samples of specimens. The possibility that the way bones were broken may help indicate the sort of creatures that did the breaking is being studied by Kay Behrensmeyer of the University of California at Santa Cruz, Pat Shipman of New York University and other investigators. For example, bones smashed to bits were probably the work of hominids getting at the marrow inside, a technique not used by hyenas and other nonhominid predators. Walker is obtaining some clues to hominid diets from scanning electron micrographs made of scratches and wear patterns on teeth, while David Gantt of Washington University in St. Louis is using the scanning electron microscope in an effort to distinguish hominid from simian teeth.

The new picture of mankind emerging involves an extremely subtle transformation, which can be seen in the fossil record and described in broad outline but which still eludes full explanation. In the beginning it was a relatively straightforward dietary adaptation to open country, a widening world of scattered trees and tall grasses. The changes that gave rise to the first members of the family of man were predominantly mechanical, the development of jaws and teeth into a modified chewing apparatus powerful and efficient enough to handle an increasing proportion of tough foods.

But something about life out there in the open started selecting for bigger, more complex brains. Survival on the savanna meant competing with other animals for food, with vast browsing and grazing herds for plant foods and, on a larger and larger scale, with lions and other established predators for meat on the hoof. There was a special premium on memory capacity and planning, on cooperative defenses and on cooperative hunting and gathering, and especially on communicating. The objective of continuing research is to understand how these and other forces produced in us not only a new species, but a new kind of evolution as well. •

Research discussed in this article is supported in large measure by the Anthropology program of the National Science Foundation.

The Emergence of Homo sapiens

by Boyce Rensberger

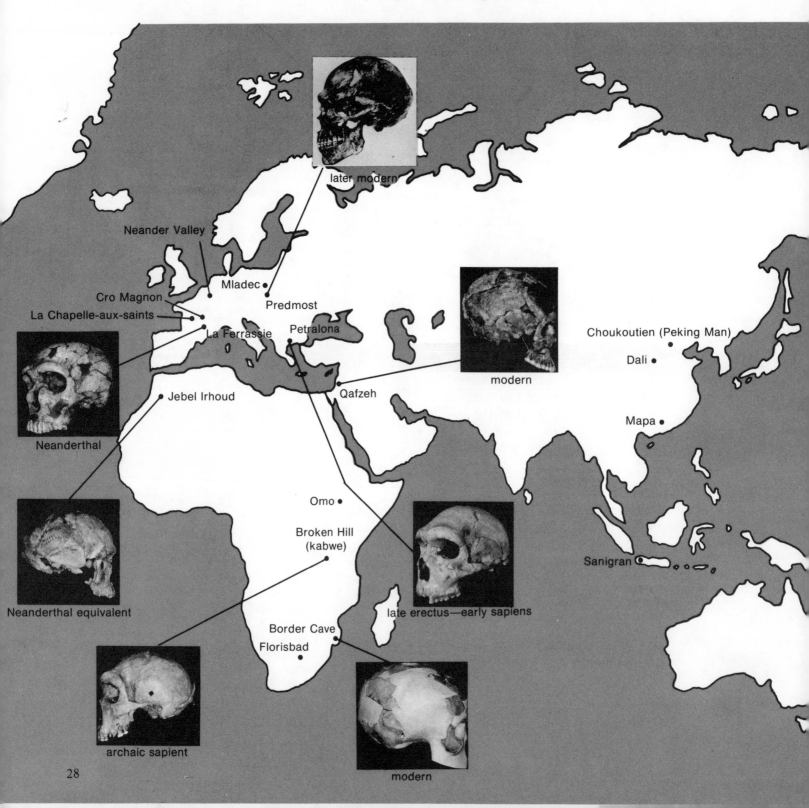

later modern

Neander Valley

Mladec

Cro Magnon

Predmost

La Chapelle-aux-saints

Petralona

La Ferrassie

modern

Neanderthal

Jebel Irhoud

Qafzeh

Choukoutien (Peking Man)

Dali

Mapa

Omo

Broken Hill
(kabwe)

Neanderthal equivalent

Sanigran

late erectus—early sapiens

Border Cave

Florisbad

archaic sapient

modern

Modern humans—*Homo sapiens sapiens*—may have evolved independently on three continents. Were the Neanderthals—*Homo sapiens neanderthalensis*—absorbed, superseded or bypassed by the wave that was ultimately to become us?

According to one side of paleoanthropology's most enduring controversy, a confrontation that was to become humanity's most dramatic clash of two cultures took place in Europe's Upper Paleolithic period, some 35,000 years ago. On the one hand there were the Neanderthals, ostensibly beetle-browed hulks who trudged about Europe and the Middle East for more than 30 millennia, starting at least 70,000 years ago. Massively built people, they purportedly made up in brute strength what they lacked in wit and cunning.

At the same time, somewhere beyond Neanderthal's mostly European range—perhaps in Africa or Southwest Asia—a new breed of human being was on the rise, fully modern, anatomically and intellectually indistinguishable from ourselves. These people would soon produce magnificent cave paintings and sculpture. The most durable evidence of their culture is sophisticated weaponry indicative of a hunting prowess well beyond that of the Neanderthals.

In a geological moment, as some interpret the stratigraphy of excavations all over Europe, the Neanderthals disappeared. They were succeeded instantly by the moderns, sometimes called Cro-Magnon after the French site where several specimens were found. By 35,000 BP (before the present) the Neanderthals were gone. Fully modern people were the sole surviving form of human being on earth.

Where did the Cro-Magnons and their ilk come from? Did they exterminate the native Neanderthals? Did they simply outcompete them for food and other resources? Did they interbreed, producing genetic mixtures that survive as today's Europeans? Or was there no sharp division at all but rather a gentle

On the line to *Homo*. Skull fragments from sites around the world suggest at least two possible routes toward the emergence of *Homo sapiens sapiens*.

Milford Wolpoff, by permission.

evolutionary blurring as one form of creature developed naturally into another more suited to changing times?

The replacement hypothesis—the notion that a group from outside invaded the Neanderthals' range and superseded them through extermination, outcompeting or interbreeding—is only the most popular guess as to what actually happened to the Neanderthal. As with so many other questions in paleoanthropology, the data that must be relied upon to answer this one have vexingly limited reliability. Many fossils are not securely dated, skeletons are often only fragmentary and it is frequently unclear whether a group of skeletons from the same site represents one or many populations. Honest but widely divergent opinions are common. The evidence supporting Neanderthal replacement, for example, can also be read to suggest a unilinear hypothesis: that fully modern human beings descended directly from the Neanderthals with relatively little contribution from outside of Europe or the Middle East.

Neanderthal's place

Since the original discovery of a Neanderthal skull in the Neander Valley of Germany in 1856—the first extinct form of human being ever found—scientific thinking about the emergence of fully modern humans has been in more or less continuous ferment. The evidence is, in some ways, more obscure than that for the earlier stages of human evolution. In recent years the effort to understand the final steps in human evolution—the steps that gave rise to our own kind, to people indistinguishable from us—has been overshadowed, at least in public perception. More attention, for example, has been drawn to the search for the earliest hominids, the first, still-apelike creatures considered to be on the direct evolutionary line to *Homo*, even though kinship to us might lie in hardly more than two-legged locomotion. (For a fuller discussion of human

evolution up to the hominid divergence see the *Mosaic* special on human origins: Volume 10, Number 2.)

The earliest known creatures that are undisputedly hominids (members of the human family after it diverged from the pongid, or ape, family) date from 3.8 million years ago. They are of the species known as *Australopithecus afarensis*, a two-legged animal with a body of rather human proportions, but not much more than a meter tall and with a head only slightly different from an ape's. According to a newly emerging interpretation, *A. afarensis*, named in 1978, could have been the common ancestor of two lineages. One included the two previously known forms of *Australopithecus*: *A. africanus* and *A. robustus*. The other lineage was *Homo*. The oldest *Homo* known, usually called *Homo habilis*, is represented by the famous skull 1470 from Kenya, dated at about 1.8 million years. Its brain was only half the size of that of people living today.

By about 1.5 million years ago, *H. habilis* had evolved into *H. erectus*. This appears to have been the first hominid to spread from Africa into Eurasia. Peking Man, who may have lived anywhere from 700,000 BP to 400,000 BP, is one well-known example. Not until the final 10 percent of the 3.8 million-year span—within the last 400,000 years—did the earliest examples of *Homo sapiens* emerge. These, however, were still not fully modern people. Their brains, for one thing, were only about 83 percent as big as ours on the average. Usually called "archaic sapiens," they ranged over most of the Old World.

Only in still more recent times—perhaps around 70,000 years ago—did they evolve, in Europe, into the classic Neanderthals, who are designated *Homo sapiens neanderthalensis*. The archaic sapiens also evolved into fully modern peoples, *Homo sapiens sapiens*.

There is evidence that the evolutionary growth of the brain—a trend that began early and slowly in hominid evolution—attained its present level something like 115,000 years ago—the last 3 percent of the 3.8 million years of the hominid career. Some say it was not until the last one percent. It remains one of the great challenges of anthropology to discover which of those figures is so: by what route—through, around or over Neanderthal—archaic sapiens became modern.

Another way

While the replacement hypothesis is perhaps the most widely held answer to the challenge, advocates of the unilinear hypothesis have not been overborne. They hold that the transition in Europe from Neanderthal to fully modern people may not have been as instantaneous as is often implied. Many thousands of years can disappear into a geological instant and, advocates note, the skeletons called Neanderthal show a high degree of variation that cannot be ignored. Some Neanderthals are decidedly more modern looking than others; some specimens even appear transitional between classic Neanderthals and moderns.

After all, the differences between the Neanderthals and their modern successors are largely matters of degree. Brains are already at their maximum size. Nearly all the differences involve decreases in the robustness of bones. People become less heavily muscled and bones become correspondingly lighter and thinner. Skeletal buttressing diminishes. Once-massive brow ridges become smaller, and what is left of a snout continues to recede under the eyes and nose.

In the unilinear view there was no cultural clash, just a gradual evolution, largely confined to Europe in the case of the Neanderthals, though similar changes would have been taking place independently and perhaps, though not necessarily, coincidentally in Africa and Asia as well. Australia, like the New World, remained uninhabited until quite recently. People first reached Australia around 40,000 BP. They appear to have been fully modern. Entry into the New World is more controversial, with most estimates ranging from 12,000 to 30,000 BP. (See "Pre-Clovis Man: Sampling the Evidence," *Mosaic*, Volume 11, Number 5.)

For years the Neanderthal controversy rested at this point, with few new developments pointing either way. Adherents of neither side could point to reliably dated remains of fully modern people much older than about 35,000 years, and certainly not from anywhere outside of Europe and the Middle East.

Ancient moderns

Then a controversial new interpretation of ancient human bones—first found some 40 years ago in Border Cave in South Africa, 400 meters from the Swaziland border—suggested that modern people were living in southern Africa a startling 115,000 years ago. This was fully 45,000 years before the more primitive Neanderthals first appeared. As some view it, the replacement hypothesis received a major boost, and the unilinear hypothesis a major setback, at Border Cave.

Was Africa the real birthplace of *Homo sapiens sapiens*? Did descendants of those early Africans spread north through the Middle East to swamp the Neanderthals and become the ancestors of today's Europeans? Or did modern peoples evolve independently in Africa and Europe and, presumably, in Asia too? Those are among the questions that hang on the Border Cave dates. Unfortunately, the dates are far from secure; again, vexingly, the evidence can be read several ways.

"There's no doubt that the Border Cave specimens are fully modern," says G. Philip Rightmire of the State University of New York at Binghamton. His detailed study of the single, adult partial skull found there (the other remains are an infant skull and an adult mandible) has established that fact. Using a statistical analysis of 11 measurements of the partial skull (including such things as the projection of the brow ridges and the distances between various bony landmarks), Rightmire has established that the Border Cave *Homo* falls within the range of variation exhibited by living peoples. Fur-

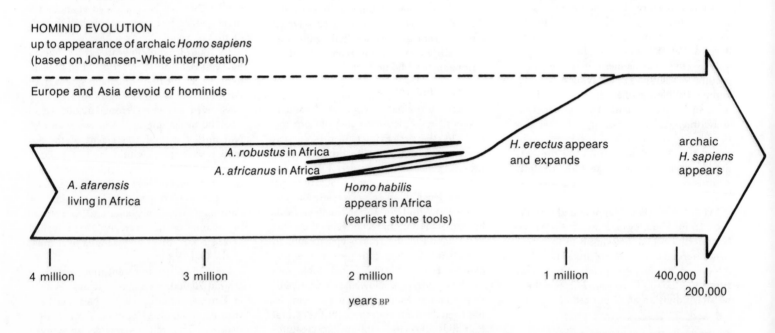

HOMINID EVOLUTION
up to appearance of archaic *Homo sapiens*
(based on Johansen-White interpretation)

Europe and Asia devoid of hominids

A. robustus in Africa
A. africanus in Africa

A. afarensis living in Africa

Homo habilis appears in Africa (earliest stone tools)

H. erectus appears and expands

archaic *H. sapiens* appears

| 4 million | 3 million | 2 million | 1 million | 400,000 |

years BP

200,000

The place of Neanderthals. Time line represents hominid evolution and two theories of the Neanderthals' place in the line to modern *Homo*: In the unilinear hypothesis, they are in the main stream; in the replacement hypothesis, they were bypassed.

ther, it comes closest to resembling today's so-called Hottentots, a South African ethnic group similar to the Bushmen (or San) but rather distinct from African Negroes. "The idea that fully modern humans appeared only 35,000 to 40,000 years ago is certainly subject to quite drastic change....

"The problem is, though, that the dating isn't that solid. There's a good deal of assumption-making going on before one can arrive at the date of 115,000 years," Rightmire observes.

The scientist principally responsible for the date, Karl W. Butzer of the University of Chicago, is rather more confident. And he sees major implications not only in such an early emergence of modern humans but in that it may have taken place in Africa. Since most anthropologists are of European ancestry, he observes, it has been almost a foregone conclusion that Europe must be the homeland of modern human beings. "Border Cave completely explodes contemporary thinking about *Homo sapiens sapiens*," says Butzer.

Assumptions and inferences

Like many other fossils and artifacts from the crucial period in human evolution between 400,000 BP, when *Homo erectus* died out, and about 35,000 BP, when modern forms become well established, the Border Cave remains are not easily datable. They

Australopithecus afarensis

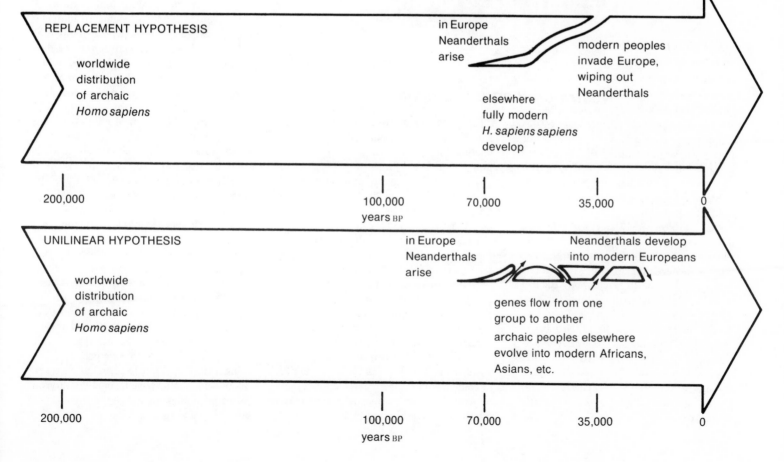

Australopithecus robustus

The younger sediments in the cave, back to one laid down about 50,000 years ago, have been radiocarbon dated. Using intervals between the radiocarbon dates, Butzer has calculated rates of sediment accumulation and extrapolated the rates to older sediments. He has also correlated the cave's cold-phase sediments with climatological data from ocean cores. By these methods, Butzer calculates that the skeletal remains at issue came from sediments deposited 115,000 years ago during a period of cool and moderately wet climate. From the bones of animal species found at the same level, it has been deduced that the habitat was then a mosaic of woodland and savanna.

The assumptions necessary to calculate a finite date in Border Cave are enough to inspire skepticism among some anthropologists, although such methods are commonly relied upon after expression of certain caveats. At Border Cave, one additional caveat is that the critical adult skull did not come from a controlled excavation. It was found in a dump outside the cave, having been tossed there in 1940 by someone digging in the cave for guano.

An important link between the skull and the 115,000-year-old layer is that bits of sediment wedged into cracks in the bone match most closely the sediments of that layer. Even more important, in Butzer's view, was a 1941 excavation that *did* carefully document an infant skull in the same beds. And in 1974 a fully modern adult jaw was excavated from a layer estimated at 90,000 BP. It was well below the layer radiocarbon-dated to more than 50,000 years. "Dating of the key fossils to between 90,000 and 115,000 years is not proved beyond a reasonable doubt," Butzer concedes, "but it's very probable. The probabilities of being mistaken are very small."

In Butzer's view, anatomically modern people probably originated in southern Africa some time before 115,000 BP. The area meets certain geographic criteria: that evolution is thought to take place chiefly on the periphery of a species' range and that in such locales environments are often different enough from those at the core of the range, so that different traits are favored by natural selection. Southern Africa, at one extreme of the hominid range, which included much of Africa and Eurasia, would seem an ideal site. It had the added advantage, Butzer suggests, of offering a wide variety of habitats within a small area. The range, from seacoast to plains to desert to mountains, should have favored the survival of people with a high degree of intelligence and adaptability.

are too old for such reliable standbys as radiocarbon dating and not suitable for potassium/argon dating, which requires volcanic minerals. One new method, amino acid racemization, has yielded a date at Border Cave that supports the 115,000-year estimate. The technique, however, is controversial and not widely accepted. Another, molecular evolution, requires soft tissue. More sensitive methods of radiocarbon dating are in development and within a few years may be able to reach about 100,000 years or more. (For more on these subjects, see "Pre-Clovis Man; Sampling the Evidence," in *Mosaic*, Volume 11, Number 5; "Molecular Evolution; A Quantifiable Contribution," *Mosaic*, Volume 10, Number 2;

"The Significance of Flightless Birds," *Mosaic*, Volume 11, Number 3; and "Extending Radiocarbon Dating," *Mosaic*, Volume 9, Number 6.)

The chain of assumptions and inferences necessary to reach any date at all for Border Cave is typical of the problems anthropologists face at many of the key sites that bear on this crucial stage of human evolution. Butzer's Border Cave date is based on a detailed analysis of sedimentary deposits in the cave. There are some 20 layers of dust, grit, rubble and the detritus of human occupation. Each layer has distinctive geological and chemical attributes. Some, for example, contain extensive amounts of rock particles that flaked off the cave roof because of frost weathering. These layers indicate a period of colder climate. Other layers show certain mineral transformations that require protracted warm and humid periods.

Rensberger, a former science correspondent for the New York Times, *writes frequently on anthropology.*

Archaic sapients

Locality aside, there is general agreement among paleoanthropologists that *Homo erectus* was the ancestor of all later forms of human beings. From about 1.5 million years ago until perhaps 400,000 years ago *Homo erectus* was the sole human species on the planet. Specimens are known from many parts of Africa, Europe, China and Indonesia.

The transition from *Homo erectus* to *Homo sapiens* could have taken place anywhere in this vast range. Fossil skulls with features that seem intermediate between *Homo erectus* and modern people—those usually termed archaic sapients—have been found in Europe, Asia and Africa.

The best known African specimen of archaic *Homo sapiens* was once called Rhodesian man. It is a remarkably complete skull that was found in 1921 near what was then Broken Hill in Northern Rhodesia and is today Kabwe, Zambia. It was once estimated to be 40,000 years old; newer evidence, putting it in line as a possible ancestor to the Border Cave people, suggests it is at least 125,000 years old and perhaps much older.

Butzer believes the case for a southern African origin of modern human beings is now strong. From there, the Chicago researcher suggests, this evolutionary trend toward more modern features gradually spread northward, reaching the Middle East by about 50,000 BP. This date is based on some fairly modern-looking human remains from Qafzeh, in Israel. The Qafzeh bones have proved difficult to date (various methods have yielded widely differing dates), but a reasonable compromise puts the bones at around 50,000 BP. From the Middle East, gateway from Africa to Eurasia, Butzer speculates, modern peoples spread out, to replace the Neanderthals and their ilk.

Strong dissent

While several American, British and South African anthropologists tend to agree with Rightmire and Butzer about the significance of the Border Cave, there are prominent dissenters. Among them is Richard G. Klein of the University of Chicago. He has specialized in interpreting the hunting skills of peoples living over the last 130,000 years, especially in southern Africa.

"Those Border Cave remains didn't come out of excavations. They came out of dumps," says Klein, recalling the guano diggers churning through the cave deposits. (They never did find any guano.) "To me that's not evidence. I remain to be convinced that the bones are as old as they say. We've all too often been misled by this kind of thing."

Australopithecus africanus

Like many paleoanthropologists dealing with what they consider to be equivocal or isolated pieces of evidence, Klein prefers to set this one aside. It is better, he feels, to try to make sense of unarguable data. In Klein's view this approach leaves the title of oldest anatomically modern *Homo sapiens* with the Qafzeh people in Israel, if one accepts an age for them of around 50,000 years. (Various methods have given dates from 33,000 to 56,000 BP.)

Klein believes there was a replacement of Neanderthals by modern people, but that the Middle East probably makes a better candidate place of origin than does Africa. His analysis of European sites suggests that, while the replacement in any one place may have been rapid, it took some 5,000 years (from 40,000 to 35,000 BP) for the wave of replacement to sweep from the Middle East westward to the Atlantic.

Among advocates of the replacement hypothesis there is debate about whether the invaders slaughtered the natives or interbred. Most suspect both and argue about the ratio. Klein, however, takes an extreme position, rejecting flatly the notion of interbreeding: "I would think that the behavioral gulf between these two very different kinds of people would have been so great that there would have been no desire at all to mate."

Klein remains unconvinced, for example, that the Neanderthals, along with other archaic *Homo sapiens*, had crossed the mental threshhold that makes modern peoples distinctive. He disputes the contentions of other scientists that Neanderthals, who produced no art, buried their dead with grave goods. (See "On the Emergence of Language," *Mosaic*, Volume 10, Number 2.) More important, Klein argues that the

Homo habilis

cupied by their evolutionary predecessors. Later sites also show abundant remains of fish and flying birds, species absent from earlier sites. Since fishing and fowling require specialized tools and skills, Klein suspects these findings help differentiate the mental abilities of archaic and modern *Homo sapiens* on any continent.

New skills

Once the transition from Neanderthal to modern occurred, whether by competition, breeding or succession, there appears to have been a great population explosion. It has been estimated that the density of post-Neanderthal humans was anywhere from 10 to 100 times that of Neanderthals. Erik Trinkhaus of Harvard has suggested that one reason may have been the Neanderthals' greater need for food energy. He estimates that, on the basis of the massiveness of Neanderthal skeletons and the necessary corresponding musculature, they may have needed twice as many calories to stay active as their more slender successors. This, however, would account for only part of the population difference. Klein argues that, since the basic resources available to both groups were the same, modern people could have been so numerous only if they were more effective exploiters of their environment.

"I don't know what it was," Klein says, "but the people who appeared 35,000 years ago knew how to do an awful lot of things their predecessors didn't. Something quite extraordinary must have happened in the organization of the brain."

It could not have been an increase in brain size, for Neanderthal brains were already just as large as ours today. Much of the older literature, in fact, asserts that they were larger, though this is now thought to be the result of early Neanderthal samples that included mostly males. Even among people today male brains are, on the average, considerably larger than female brains with, obviously, no difference in intellectual power.

"I'm quite convinced," Klein says, "that in Europe it was a physical replacement of one kind by another. And I'm prepared to bet that that's what happened in Africa too and at about the same time."

The unilinear view

Milford H. Wolpoff would take that bet. Wolpoff, at the University of Michigan, is one of the leading advocates of the unilinear hypothesis—the view that there was no sudden, single replacement of one kind by another. Rather, he suggests, the Neanderthals by and large evolved into today's Europeans. The anatomically modern population represented at Qafzeh did not in-

Neanderthals were unable to make superior weapons. They were "rotten hunters," he declares.

Additionally, from his studies of South African sites where the bones of prey animals were preserved, Klein has deduced that people of the African Middle Stone Age, who were culturally comparable to the European Neanderthal of the period called the Middle Paleolithic, were able to bring down only the weakest and least dangerous animals. Using fossil teeth to determine the maturity of the prey species, Klein has proposed that Middle Stone Age hunters generally killed animals under a year old. Very few animals in their prime are represented in the preserved garbage of those times. The prey-age distribution is comparable to that of lions. The two exceptions are the eland and the bastard hartebeest. Unlike other bovids, both can be driven in herds. Klein

suspects that Middle Stone Age hunters learned this and drove entire herds off cliffs.

Fully modern people from the Later Stone Age, comparable to Europe's Upper Paleolithic, Klein has found, were able to kill any animal they chose. Bones from such sites reflect an age distribution closer to that of living groupings and also include remains of more dangerous animals such as wild pigs.

The difference, Klein suspects, was in the weaponry. Armed with little more than rocks and clubs, neither the Middle Stone Age *Homo* nor the Neanderthal could get close enough to an animal in its prime and they dared not approach dangerous prey. They lacked the ability to invent such superior weapons as the spear thrower or the bow and arrow that make it possible to kill from a distance. The remains of such weapons have been found in sites of modern peoples in Africa and Eurasia, but not in sites oc-

vade Europe but, instead, having derived from an archaic *Homo sapiens* there, gave rise to today's Middle Easterners and North Africans. The Border Cave people, whatever their age, then would be the ancestors of today's southern Africans. Other fossil remains from Asia, such as the Neanderthal-like people represented at Mapa and Dali in China, are in the line that led to modern Asians.

"Any theory of human evolution," Wolpoff notes, "has got to account for the differences among modern populations. A modern European skull looks different from a modern African skull. And both of them look different from a modern Chinese or a modern Australian."

Indeed, while all living peoples unquestionably belong to the subspecies *Homo sapians sapiens*, most members of each population—sometimes designated a race or ethnic group—share certain distinctive skeletal features. In fact, using the kind of statistical comparison of measurements that Rightmire applied to the Border Cave skulls, it is often possible to distinguish between rather closely related groups such as the Bushmen and the Hottentots.

"You look at what the distinctive features of modern Europeans are and then you look at the fossil populations to see where those features first appear, and you find them in the Neanderthals," Wolpoff says. One feature he likes to cite is the big nose. European anthropologists tend to euphemize the feature, including it in what they call the "mid-facial prominence," but it is clear that people of other races find Europeans distinctive because, among other things, of their noses. Europeans have the most prominent noses of the living races. It begins jutting out at a fairly sharp angle just below the brow. In Africans and Asians, the nasal bone descends well below the browline before curving outward. The fleshy part of the nose may be broader in some groups but it rarely protrudes farther than or begins to protrude as high as the Europeans'.

Neanderthals had big noses. Only half jokingly, Wolpoff says their noses must have resembled that typified by Charles de Gaulle. In Neanderthals the feature is often considered an adaptation to a cold climate, because a larger nose is presumed better for warming inhaled air. Anatomically-modern fossil populations from outside Europe, such as the people of Qafzeh or of Border Cave, lack this feature.

Wolpoff cites a variety of other anatomical features that, in the same way, are characteristic of a modern race and that first appear in the archaic *Homo sapiens* fossils from the same area. These include various

Homo sapiens (archaic)

subtle contours of the skull bones: for example the more flattened face of Asian peoples and the slight bulge that bridges the brow ridges above the nasal root in the African skull.

"To me it makes the most sense to assume that those distinctive features were inherited from the people who were already living in the area and who already had the feature," Wolpoff says. For most other parts of the world, most anthropologists accept such parsimony, he declares, but not for Europe.

Neanderthal types

One of the chief reasons, in Wolpoff's unilinear view, that the Neanderthal controversy continues is too great a reliance on typological thinking. In other words, when people think of the Neanderthals, one particular skull—often beetle-browed—or a closely related group of skulls comes to mind.

And when people think of more modern successors, they think of another set of distinctive skeletal traits, including less prominent, loftier brows. Between the two stereotypes there are great differences, and they lead to the view that the earlier could not have given rise to the later in so brief a time.

"People forget just how much variation there is in every population," Wolpoff observes, pulling, as he talks, various casts from cabinets in his laboratory and arranging them on a table. (He maintains what is considered to be the most complete collection of fossil hominid casts in the United States.) "Every feature that is considered to distinguish modern Europeans from Neanderthals can be found in [one or another] Neanderthal sample."

Modern features are rare among Neanderthals and certainly not typical, he concedes, but this is exactly what would be

Homo erectus

expected of evolution. Natural selection works by acting on traits that are already expressed. A trait may be represented at a very low frequency in a given population; if it becomes advantageous, after many generations it will come to predominate.

The Neanderthals, in Wolpoff's view, were far from homogeneous either at any one time or throughout the 35,000 or so years they existed. Modern traits—less massive bones or higher foreheads, for example—are present but rare in the earliest specimens. In the later Neanderthal populations, such traits become more common. There are even some skulls that appear to be a blend of Neanderthal and modern features, so much so that some authorities have guessed them to be hybrids. Unilinear advocates, on the other hand, see them as evidence of evolutionary transition.

Even *Homo erectus*, the immediate ances-

tor of *neanderthalensis* and other *Homo sapiens*, and often said to have been remarkably stable in its million-year career, actually changed with time. Brain size, for example, grew some 20 to 25 percent between the earliest and latest specimens.

And the transition from *H. erectus* to *H. sapiens* was gradual. There is no generally accepted way to define the boundary. There are specimens that look like hybrids of the two types and might have been taken for such if they were not dated to about 400,000 BP, when the transition was in progress. There are similarly gradual transitions elsewhere in human evolution. There are, for example, specimens that look intermediate between the archaic sapiens and classic Neanderthals. And there are, among the fossils called "anatomically modern," many examples that are considerably more archaic in appearance than are living people.

Again, because of the lack of reliable dates for many of the specimens, it is not always possible to arrange them in chronological order. But by using estimated dates and archaeological associations, it is possible to produce what amounts to a morphological continuum from *H. erectus* to *H. sapiens sapiens* into which *H. sapiens neaderthalensis* fits nicely.

Wolpoff holds that the Neanderthals were simply European representatives of a phase of human evolution through which people also evolved in Asia and Africa. This has sometimes been misunderstood as an assertion that archaic sapiens from Asia and Africa were Neanderthals. Rather simplified, Wolpoff's idea is this: Since the parent stock of all modern peoples was *Homo erectus*, and since modern people today despite minor differences all differ from *Homo erectus* in the same way, people everywhere had to evolve through intermediate stages that exhibit similar intermediate features.

These intermediate features, along with the results of natural selection in the unique European environment, in Europe produced a classic Neanderthal. In Africa, the same gradation is represented by specimens found in Ethiopia's Omo Valley, at Florisbad in South Africa and at Jebel Irhoud in Morocco. They lack certain distinctive Neanderthal features; instead, they have uniquely African traits. Comparable Asian specimens would be the skulls from Mapa and Dali in China.

Mainstream Neanderthals

Wolpoff also asserts that the Neanderthals were not the dull-witted brutes that Klein envisions. In fact, he sees no reason to doubt that they were anything other than squarely on the intellectual continuum, almost if not already the equal of modern human beings.

One recent discovery in France lends new support to this view. Bernard Vandermeersch of the University of Paris has found a Neanderthal skeleton in clear and direct association with stone tools more sophisticated than the Mousterian tools that are typical of most Neanderthals. These advanced tools are of a type known as Chatelperronian. The kit includes such Mousterian examples as scrapers and irregularly shaped flakes for cutting. But it also includes some of the long, regularly shaped blades, struck from a flint core, that are typical of the tool kit of more modern people.

Until now the finer, Chatelperronian tools have always been considered early examples of the work of modern people. Now it appears that Neanderthals were capable of just that transition to more advanced technologies. Additionally, the modern people of Qafzeh, considered by replacement advocates as possible sources of the invasion, made and used the cruder Mousterian tools. "What is all comes down to," Wolpoff argues, "is that if you look at all the European evidence, there is no great jump. You don't need invasions."

But are there inconsistencies? Would the people of Border Cave, assuming they were fully modern 115,000 years ago, have bided their time in Africa while less advanced peoples occupied Eurasia? Wolpoff is reserving his opinion on the reliability of the Border Cave date. But, he argues, it makes little difference how old those people are. Citing Rightmire's conclusion that they most closely resemble modern Hottentots, Wolpoff suggests that they were simply the ancestors of today's southern Africans. The distinctively African features in the Border Cave skeletons do not appear in any European fossils. From this, Wolpoff concludes that the Border Cave people are unlikely to have contributed in any large part to modern European ancestry.

The unilinear hypothesis should not be understood to rule out mating between otherwise separated groups. Indeed, most authorities assume it must have been a common occurrence. It is the norm today in many traditional cultures for men and women to seek their mates from other bands or clans or villages. This practice, if extended indefinitely, means that genes are flowing more or less continuously over the entire inhabited range. One effect of this practice, well documented for living peoples, is that physical traits that are predominant in one area slowly diffuse to the surrounding areas. If a trait is advantageous in all environments, it will quickly spread. But if the environment of the surrounding area does not favor the trait, the introduced gene will remain at a low frequency. If bearers of this gene in the surrounding area chance to mate with someone from an area still more distant from the trait's center, the gene will be spread farther but still at a frequency related to its utility.

Shared traits

Many physical traits among modern peoples (skin color, height, head shape, etc.) are distributed in this way and will continue to exist in continua so long as there is outbreeding at the range margins. So long as the environment at the core of the area exhibiting the trait continues to favor that trait, it should remain common there. Like ripples on a pond, the trait should continue spreading so long as the force making the

Homo sapiens neanderthalensis

ripples remains active.

In this way, Wolpoff suggests, traits that are only locally advantageous will spread some distance away but will remain rare at that distance. On the other hand, traits that are advantageous in all environments, such as a larger brain, will spread throughout the inhabited region and reach high frequencies throughout.

The flow of universally advantageous genes, Wolpoff suggests, would be likely to spread them to neighboring peoples before the originating population progressed so far that it could use the advantage to invade or exterminate its neighbors.

Replacement advocates, of course, disagree. They envision early peoples as so widely dispersed that, from time to time, groups became cut off—perhaps isolated by a desert or a mountain range. These insular groups would not spread any of their newly

evolved advantages until they had developed well beyond their contemporaries elsewhere and then breached the isolating constraints. Thus big-brained peoples, if isolated long enough, might eventually break out and replace their small-brained contemporaries.

There is no clear sign that the controversy over the emergence of anatomically modern peoples will be resolved soon. Undoubtedly more fossils will be found. But perhaps more important, existing discoveries must be reanalyzed with the aid of new techniques, new or extended dating methods and fresh eyes. One very serious handicap to any single investigator is the difficulty of access to most of the original fossils (which are housed in isolated collections around the world) or even to casts of the bones (which are either expensive or unavailable). Hominid fossils are often treated as the personal property of their discoverer and sometimes access is

Homo sapiens sapiens (Cro Magnon)

granted only to a favored few. A full description of the bones may be years or decades in coming, and until then convention dictates that no one else may analyze or interpret the material in detail.

Compared to the rich and active lives led by thousands or millions of members of now extinct hominid species in the many past environments of the planet, the bits of bone that have been found in the past century—in all only a few score—are a pitifully meager basis from which to develop a believable story of human evolution. Still the broad outline of a fairly coherent story has emerged. Indeed, the origin of human beings in apelike ancestors is among the best documented speciation events in paleontology. Only the details remain troublesome.

And as the details come closer to illuminating the differences and similarities among living peoples, we may rightly become more rigorous and, inevitably, more contentious in evaluating the evidence. Clues to many of the most important events or processes in

human evolution may never amount to proof, at least in the eyes of other disciplines.

And yet it is nothing less than the heritage of our species that is at stake. We have, after all, come a long way from the view of the shocked lady who is alleged to have said, when Darwin's ideas of descent from the animals first burst forth, "Let us hope that it is not true, but if it is, let us pray it does not become generally known."

Like orphans searching for our parents, we want to know where we came from, how we got here, how we are really related to the rest of the living world. It can be argued, in fact, that providing this knowledge is paleoanthropology's highest use. In this light, even the smallest quibbles about how human evolution took place are matters of vital substance for us all. •

The National Science Foundation contributes to the support of research discussed in this article through its Anthropology Program.

Pre-Clovis Man: Sampling the Evidence

by Ben Patrusky

The probable, the possible and the provable aren't all the same where the earliest inhabitants of North America are concerned.

No serious archaeologist argues about the origins of the first human inhabitants of the Western Hemisphere or about how they arrived. These paleo-Indians strode—or drifted—in, from what is now Siberia across the Bering Strait, over a land bridge that joined the Old World to the New during the last great ice age. They came as big-game hunters on the trail of moving herds of giant elephants and other megafauna of the Pleistocene epoch. They were first occupants of Beringia—a continent-sized landmass linking Asia and North America—which lured them as it lured their game. Then shifting needs and opportunities propelled them farther east and south.

Not much of this is in question. But one big question remains: When did they come; how early were early humans in the New World? Since the early days of the 20th century, New World archaeologists have been debating the issue—often heatedly. Timetables abound. Robert L. Humphrey of George Washington University has dubbed the ongoing controversy "The Hollywood Complex, or my early man site is earlier than your early man site."

The heat of the debate doesn't surprise Dennis Stanford, director of paleo-Indian archaeology at the Smithsonian Institution. "People digging at the roots of America's ancestry tend to be highly messianic," he explains. "They've invested time, money, thought and often reputation. Sure they want their labors and finds to prove significant. Sometimes they're not very careful about interpreting their results."

The debate about humankind's dawning in the New World is far from settled. But in recent years highly trained archaeologists have begun to accumulate impressive bits of evidence that suggest migrations at a time much farther back than might have been accepted only a few years ago. Some human occupancy as much as 20,000 or 25,000 years ago is beginning to look, if not feasible, at least arguable. Dates much older than that, however, while postulated, still strike more heat than light from among the disputants. At the present stage of knowledge and of dating technology, there is slim chance for the dispute over Western humankind's antiquity soon to be settled. But a critical

Tracking origins. Fossilized tracks of a mammoth lead to the site of its remains at Murray Springs, Arizona.

look at the evidence and the inferences to be drawn from it can be highly suggestive.

Clovis and before

The pivotal point of the debate is on the order of 12,000 years ago—or 12,000 BP, for "before the present," in the parlance of the professionals. That represents the earliest totally accepted date for human appearance in the New World.

Evidence of human presence then is incontrovertible. It takes the form of a special kind of artifact: a man-made tool, a projectile point with a highly distinctive shape, a manufactured weapon for killing mammoths and other big game. The points are bifacially fluted; longitudinal flakes have been chipped from the base on both sides to form grooves. With fluting, the point could be attached to a wooden shaft for use as a spear or dart. These fluted points have been found at sites ranging from the Pacific coast to the Atlantic coast of North America and from Alaska to central Mexico. In the sites where carbon 14 dating is unequivocal, all the projectile points have been found to date within a very narrow window of time—11,500 to 11,000 BP.

Because the points were first found in abundance in a locale called Clovis in Blackwater Draw, New Mexico, the points have been designated Clovis points—and their manufacturers the Clovis Culture. The Clovis Culture is also often referred to by another name: the Llano Complex, for the Llano High Plains region of the Southwest. The designation embraces not just the fluted projectile points but the entire bone and stone "tool kit" associated with them.

But were the bearers of the Clovis Culture the first inhabitants of the New World? Or were there earlier bands of migrant-hunters whose groping cultural evolution ultimately gave rise to the technologically advanced Clovis? That question is at the heart of the peopling of America debate.

That the Clovis people were the very first Americans is a position promulgated by Paul S. Martin of the University of Arizona. He has introduced the "overkill theory" to support his arguments. Martin contends that the Clovis people, already skilled big-game hunters, swept into the Americas from Alaska in a single, rapid migration about 12,000 years ago. There they encountered a hunter's paradise: a land teeming with mastodons and mammoths, sloths, giant cats, horses, camels and bison. The archaeological record shows that many of these species became

extinct coincidentally with the advent of Clovis. Clovis, according to Martin, cut a rapacious swath through the hemisphere, exterminating these Ice Age mammals.

Other New World archaeologists contest the overkill hypothesis. In their view dramatic changes in climate and vegetation, in the wake of a significant glacial retreat, were the agents of extinction. More to the point, a number of these investigators are convinced that humans trod the soil of the New World far in advance of Clovis's distinctive appearance. Says Richard S. MacNeish, director of the Robert S. Peabody Foundation for Archeology: "There can be little doubt that man was here well before 12,000 years ago, as we have about 30 sites with more than 2,000 recognizable artifacts and with more than 60 radiocarbon determinations before 10,000 B.C.E. (Before the Common Era)." How much earlier than 12,000 years ago? MacNeish suggests that humans "may have first crossed the Bering Strait land bridge into the Western Hemisphere between 40,000 and 100,000 years ago."

The pre-Llano vista

One of the most prominent figures in the controversy over New World habitation is anthropologist and geoscientist Vance Haynes, also of the University of Arizona. To Haynes, a self-styled "archaeological conservative" who has been stalking Clovis for more than two decades, has been delegated the *ad hoc* role of both arbiter and devil's advocate in the assessing and validating of evidence presumptive of prehistoric migrations into the New World. Haynes's judgment: "There is no one place where the evidence (for pre-Llano cultures) is so compelling that if you looked at it in a court of law you would want to be tried on the basis of that evidence."

The evidence to date for a pre-Llano presence in the New World is not yet "airtight," admits the Smithsonian's Dennis Stanford. "But some of what we do have looks really good and very compelling and can't be lightly dismissed."

Some of the key pieces of evidence of the existence of a pre-Llano people include:

• **Pikimachay Cave in Highland Peru,** where MacNeish has uncovered artifacts he insists date back as much as 21,000 to 25,000 years. "My excavations of Pikimachay Cave have proved to me that pre-10,000 B.C.E. and pre-20,000 B.C.E. remains of man do exist," he says unequivocally.

Peruvian cave. Excavations of Pikimachay Cave in Peru offer evidence that humans were in the Western Hemisphere before 20,000 B.C.E.

Richard S. MacNeish, by permission.

MacNeish discovered the cave in 1967. It is a huge rock shelter, 85 meters long and 25 meters deep, situated halfway up a hill stepped by ancient terraces. Meticulous excavation has revealed sequential strata of habitation, "a series of floors on which man had clearly lived," declares MacNeish. A roof-fall that occurred about 9,000 years ago securely sealed off the lower, earlier deposits from any possible intrusion or artifact contamination from strata lying above and representing a later chronology.

Beneath this rocky lid lie seven strata, showing evidence of a number of occupations by man back as far as 25,000 years. Using carbon-14 dating, University of California at Los Angeles scientists assigned BP time slots of 19,660 ± 3,000; 16,050 ± 1,200; 14,700 ± 1,400, and 14,150 ± 180 to the upper four strata, respectively, though the lowest has not been so dated. The antiquity of the earliest dated level, MacNeish reports, is confirmed by an independent analysis by Isotopes, Inc., which reported an age for that stratum of 20,200 ± 1,000.

All told, says MacNeish, almost 300 "indisputable" artifacts were unearthed in association with more than 800 bones of sloths and other extinct Ice Age mammals. Perhaps the best evidence was turned up at the upper level, dated at 14,150 BP, which was found to contain 133 artifacts, including stone projectile points and scrapers. Many of the recovered artifacts are very crude, says MacNeish, and a far cry from the sophistication of Clovis, suggesting that they were the tools of unspecialized hunters and gatherers.

Mr. Patrusky, who writes most often on biomedical subjects, was the author of "Molecular Evolution; a Quantifiable Contribution," in Mosaic, *Volume 10, Number 2.*

• **Valsequillo Basin, Puebla, Mexico,** where in the early 1960s investigators stumbled upon a trove of extinct-animal bones as well as some crude stone artifacts along a dry river bed. Subsequent stratum-by-stratum excavation turned up an ever receding chronology of hunters slaughtering Ice Age animals. At about 30 meters down, at a level dated by carbon-14 at 22,000 years, archaeologist Cynthia Irwin Williams found evidence of a mastodon dismemberment by ancient butchers.

• **Tlapacoya, Valley of Mexico, near Valsequillo,** where Mexican investigators Jose Lorenzo and Lorena Mirambell found bones in association with a shallow depression containing charcoal that yielded a carbon-14 date of 24,000 BP (give or take 4,000 years) and presumed to be the remains of an ancient hearth. Moreover, the investigators uncovered some obsidian artifacts, including a curved blade found buried under a large tree trunk, that gave the same 24,000 BP date.

• **Del Mar in Southern California.** A group of 11 human skeletons has been found along the California coast and dated by Jeffrey Bada of the University of California at San Diego, using a technique called amino acid racemization. It is based on the assumption that amino acids in protein undergo a configurational change—from the so-called *L* or left-handed sort to the *D* or right-handed kind—as bone fossilizes, and that this switching goes on at a measurable, clocklike rate until the sample reaches half-and-half equilibrium. The oldest date determined on a skull found at Del Mar gave a racemization number of 48,000 BP. Two other samples—dated at 23,000 and 17,150 BP—have been confirmed by radiocarbon dating on bone collagen performed by the University of California at Los Angeles' Rainer Berger.

• **Santa Rosa, California,** an island 25 miles off the coast of Santa Barbara, where Berger and Phillip Orr, curator of the Santa Barbara Museum, discovered a red burn area about three meters in diameter in association with stone tools and the bones of dwarf mammoths. The supposition: that the burn area actually had served as a hearth where paleo-Indians cooked the horse-sized mammoths as far back as 40,000 years ago. Carbon-14 dates derived from bits of charcoal found in the firepit suggest the age. According to Berger, during the late Pleistocene (100,000—10,000 years ago) Santa Rosa and three other nearby islands may have formed a single landmass that was joined to the mainland via a narrow neck of land across which foraging dwarf mammoths and their human predators could readily wander.

Chemical analysis. Pat Masters (left) and Jeffrey L. Bada confer in the amino acid racemization lab at Scripps Institution of Oceanography.

Kerby Smith, by permission.

• **Meadowcroft Rockshelter, Pennsylvania.** On a sandstone outcrop 65 kilometers southwest of Pittsburgh, James M. Adovasio of the University of Pittsburgh has unearthed an assemblage of unifacial tools associated with radiocarbon ages of 13,250, 14,850 and 15,120 years. One especially significant find was a bifacial projectile point—"like a fluted point except that it has no fluting"—that dates earlier than the Clovis point and may be ancestral to it. (A detailed discussion of paleo-Indian sites in Pennsylvania and Virginia appeared in *Mosaic*, Volume 8, Number 2.)

• **The Selby and Dutton sites in eastern Colorado.** In 1975 pond-dredging crews on farms near Wray, Colorado, unearthed two Pleistocene fossil sites. Alerted to the discovery, a Smithsonian Institution archaeological team that happened to be working nearby raced to the scene. In subsequent (and as yet far from complete) excavations, under the direction of Dennis Stanford, the team initially discovered fluted projectile points and other manifestations of Clovis. Digging deeper, they found evidence of the presence of earlier cultures that included what are called bone expediency tools. These are manufactured by producing a spiral fracture in bone and using the sharp edges for butchering or hide-working. Bone flakes, presumably the waste or debitage from tool production, and bone, apparently processed for the removal of marrow, were also found. The tools, made from the bones of Ice Age horse, bison, mammoth and camel, were found in "chronologically secure" layers extending back a possible 20,000 years. What makes the site especially exciting, according to Robert L. Humphrey, is that it "reveals the only evidence to date of an archaeological culture stratigraphically *in situ* below a level containing Clovis fluted points and other manifestations of the Llano Complex."

• **Old Crow Basin, Yukon Territory,** where in 1966 was discovered a caribou leg bone that had clearly been worked by human hands to produce a tool called a flesher—a back-scratcher-like implement used to scrape animal hides. William Irving of the University of Toronto and Richard Morlan of the National Museum of Man, in Ottawa, who now direct major archaeological research projects in the basin, established a carbon-14 date for the flesher of 27,000 BP. Since then, other worked-bone specimens have been found and dated as far back as 41,000 years.

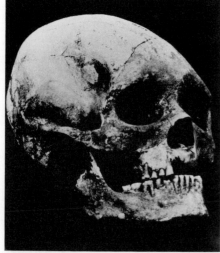

Ancestral skull? This paleo-Indian skull was dated at nearly 50,000 years old through a still-controversial amino-acid dating process developed by Jeffrey L. Bada.

San Diego Museum of Man, by permission.

Not foolproof

What says Clovis specialist Haynes to all this? "Tantalizing but not foolproof. . . .I am a skeptic. I remain a skeptic. I have yet to see *unequivocal* evidence of pre-Clovis. Each (site) has some kind of uncertainty connected with it." On his list of "uncertainties," Haynes includes:

• **Skepticism about radiocarbon dates.** Samples may become contaminated by groundwater, he says, and throw the true carbon-14 date off accordingly. Case in point: the Meadowcroft site. It is possible, he says, that rainwater could have brought in dissolved carbon from nearby coal deposits, thus disturbing the precision of the radiocarbon date. Haynes isn't saying that is what happened. "It's just that we can't be sure it didn't happen." Haynes also expresses doubts about "one of the best pieces of evidence

Pushing back the date. Archaeologist William N. Irving at Old Crow Basin in the Yukon Territory. Artifacts found there are radiocarbon dated to 27,000 BP.

Kerby Smith, by permission.

around for a pre-Clovis presence in the New World," the Old Crow flesher. "There's no way to say for sure that the bone was not pulled out of the ground and worked over at a much later date," he says.

• **Suspicion about amino acid racemization.** In a thorough review of this dating technique, David von Endt, a research chemist at the Smithsonian Institution, writing in a monograph on pre-Llano cultures published in 1979 by the Anthropological Society of Washington, D.C., concludes: "I view no projected amino acid date as reliable." Reason: The rate of turnover from L- to D-form amino acids depends on a mix of complicated factors that have yet to be fully reckoned into the racemization clock.

• **Uncertain primary or cultural context.** "The primary requirement," says Haynes, "is. . .an assemblage of artifacts that are clearly the work of man. Next, this evidence must lie *in situ* within undisturbed geological deposits. . . .Lastly, the minimum age of the

site must be demonstrable by primary association with fossils of known age or with material suitable for reliable isotopic dating. These requirements have now been met repeatedly for the late paleo-Indian period (Clovis) but they have not yet been met repeatedly for earlier periods." Example: the "questionable" primary context of the Santa Rosa firepit. Haynes wonders whether this so-called hearth may not actually be the product of a natural brush fire—not an uncommon phenomenon on this chaparral-dense island.

• **Artifact versus geofact, or when is a tool a tool?** Haynes talks about a "bandwagon effect, where some archaeologists see artifacts everywhere they look." In many cases, he says, the artifacts are nothing more than geofacts or ecofacts, pseudotools produced by agents other than human. "I think if we were to dig anywhere there are Pleistocene, coarse-grained sediments or bones," says Haynes, "(we) would find something that could be interpreted as artifact. In other words, there is a sort of 'background noise' in the buried record of things that can be taken as artifacts." Case in point: Calico Hills in the Mojave Desert, near Yermo, California, where in 1968 archaeologists spotted what they proposed to be crude, very ancient flint artifacts in a massive deposit of alluvial gravels of Pleistocene age. The investigators contend that some of the stone specimens show flaking marks that could have resulted only from human intervention. But Haynes, who has examined the site and specimens on six different occasions, reads the "evidence" another way. As he puts it: "Evidence for artifacts remains uncompelling. . .(and) a natural origin cannot be precluded. In fact, normal natural processes are adequate to explain the origin of all the phenomena observed. Even the best specimens could have been chipped and flaked naturally, especially in view of the fact that each 'artifact' has been selected from literally hundreds of thousands of individual pieces of chert." Similarly, two criteria seem to distinguish bone artifacts: evidence of a spiral fracture (a fundamental step in tool production or marrow processing) and of polishing (indicative of tool use). But other forces, agents other than man, can break or polish bone—e.g., trampling and gnawing by other animals.

• **Absence of stone artifacts.** Thus far Old Crow and the Selby/Dutton sites (at strata below Clovis) have yielded only bone artifacts. No stone tools are in evidence and, according to Haynes, "you can't produce flake scars by hitting bone against bone. Further, there's little to support the validity of an all-bone culture. If there were stone tools at Dutton

Mammoth jaw. Anthropologist Vance Haynes examines a mammoth jaw at the Murray Springs Clovis site in Arizona.

Helga Teiwes/Arizona State Museum, by permission.

or Old Crow, where are they?" Haynes remains skeptical about the seven tiny stone "impact flakes" suggestive of human origin turned up by Stanford's team at Dutton and described as tailings produced as a result of impact from a chopping tool.

Counterarguments

Needless to say, Haynes's arguments breed counterarguments among those convinced of a significant pre-Llano presence. Old Crow archaeologist William Irving, for example, suggests that stone tools may not be all that essential to the support of human life. In fact, he contends, bones that could have performed all the necessary tasks of hunting, piercing, butchering, skinning and perforating have been recovered from deposits at Old Crow and elsewhere. Two years ago, Dennis Stanford gave a dramatic demonstration of bone's versatility. Using only bone tools, Stanford showed how it was possible to butcher and process the remains of an elephant, operating on one that had died of natural causes at a Boston zoo.

As for the artifact versus geofact issue, Stanford readily concedes that archaeologists may be led astray by wishful thinking—"seeing what they want to see." But, he points out, with bone, for instance, there are criteria to help the investigator with a trained eye to distinguish between tool and pseudo-tool. For example: With tools, only the ends—the working parts—tend to become polished and worn. This discriminatory polishing doesn't often happen in nature, unless only the ends have been gnawed. But then there would generally be tooth marks to help make the distinction.

Also, bones that exhibit spiral fracture by other-than-human cause are often those broken after the bone had dried or as a result of the animal's falling or twisting its leg. In both cases these bones are not likely to exhibit impact depressions suggestive of human workmanship. Moreover, says Stanford, at the Selby and Dutton sites most of the bone-artifact specimens were broken the same way; they show single and multiple points of impact located in the same area of the bone. "These suggest a pattern that cannot be attributed to the random breakage patterns expected if the bones were either broken or polished due to natural conditions," he says.

In an effort to eliminate all doubt about bone-artifact authenticity, Stanford has been overseeing a number of experimental studies aimed at producing hard, quantifiable data on bone fracturing, modification and use. From his elephant-butchering trials, for instance, he has developed information regarding tool-wear patterns. Another study has Gary Haynes, a doctoral student, feeding buffalo and horse bones to bears in the National Zoological Park in Washington, D.C., to see just how the animals break and gnaw bones. Haynes is also observing how the bones of wood bison, newly killed by wild wolves in Canada, are altered by the natural environment. In an applied offshoot, Stanford is consulting with veterinarians who are analyzing bone breaks in race horses. Are the breaks natural, or did someone break a horse's leg intentionally? The ability to discriminate has relevance for insurance companies as well as archaeologists. Stanford is hopeful that he will soon get permission to put elephant bones around a pond at the National Zoo's Conservation and Research Center near Front Royal, Virginia, to see if and how other free-to-roam animals treat the specimens.

Meanwhile, at Santa Rosa Island, museum curator Phillip Orr has built model pit barbecues to help determine whether the red-char area he and Rainer Berger discovered actually served the earliest immigrants as a hearth or whether, in fact, it was the result of a natural brush fire. His experiments produced the same kind of deep-soil burning (to a depth of about 80 centimeters) as seen in the purported pre-Llano firepit, but different from the burn pattern of a natural chaparral fire.

Whither Clovis

When Haynes, eminently known for his "insistence on methodological exactitude," drops his devil's-advocate mask, he readily professes a belief in a human presence that may far antedate Clovis, a presence he believes will become demonstrable and unquestionable in time. In fact, he now seems near-convinced about the validity of MacNeish's Pikimachay Cave finds—specifically those from the 14,150 BP level. The Valsequillo deposits, dating from 9,000 to 21,000 years ago, also enthrall him (although the picture there remains clouded by another kind of uncertainty: unsubstantiated charges that workmen may have deliberately planted some of the would-be artifacts).

But far from quashing the debate, the strong indications of a pre-Llano presence have introduced a major new wrinkle to the controversy, having to do with the origins of the Clovis Culture. Is it, as Martin, Vance

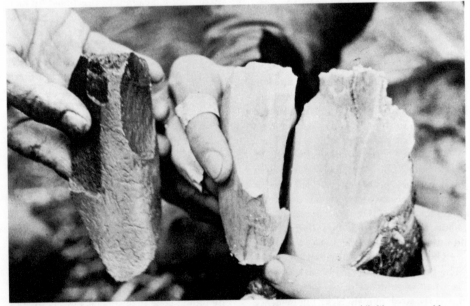

Clovis tools. A bone tool flaked from an elephant bone core is similar to a tool (left) recovered from Old Crow site.

Smithsonian Institution.

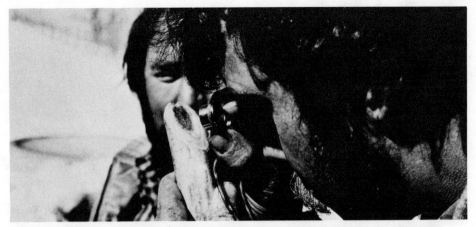

Antler tools. Dick Morlan examines an antler wedge (above) with Dennis Stanford looking on. Morlan uses an antler hammer and wedge (right) to remove a membrane from an elephant bone, facilitating bone-breakage experiments.

Chip Clark/Smithsonian Institution.

Haynes and others contend, an import by a technologically advanced people who brought their Upper Paleolithic wisdom with them from Siberia? Or is it, as MacNeish, Stanford and other pre-Llano advocates maintain, a homegrown product—an evolutionary outgrowth of already-in-place, pre-Clovis cultures? Asked another way: Is the Clovis point, that superior paleo-Indian invention that allowed Llano hunters to take full advantage of the megafauna-rich Pleistocene environment of the New World, a candidate for the first truly American patent?

In Haynes's view, even if pre-Llano hunting cultures entered the Americas more than 30,000 years ago, they did not develop into technologically skilled artisans anywhere near the caliber of Clovis. As he sees it, Clovis represents an entirely independent migratory swarm, a late-glacial sweep distinct from earlier, "inconclusive" movements into the New World. "It now appears," says Haynes, "that the *main* peopling of the New World took place during deglaciation, when something akin to a population explosion occurred between 11,000 and 11,500 years ago by mammoth hunters entering from Alaska and finding abundant and untapped resources."

Haynes offers up this scenario: During the peak of the most recent glacial epoch—anywhere from 14,000 to 20,000 years ago, he says—a large portion of the earth's ocean water was stored in Northern Hemisphere ice sheets, causing the sea level to drop by scores of meters. What emerged was the 1,600-kilometer-wide Bering land platform, which made Alaska as much a part of the Asian continent as the North American and which allowed for easy migration from Old World to New.

The bridge, however, was not a thoroughfare from the Old world to the *whole* of the New; two great icecaps—the Cordilleran on the west, and the Laurentide on the east—covered much of Canada and much of the United States. Joining as they did at the foot of the Canadian Rockies, the giant glaciers created an ice barrier, a wall to southerly migration.

The hunters remained mired in an Alaskan cul-de-sac until about 12,000 years ago, when a period of marked glacial retreat opened a north-south corridor between the two icecaps. The progenitors of Clovis, confined until then in central Alaska, swept south in pursuit of the mammoth and other quarry. Once through, they dispersed rapidly across all of North America and into Mexico. All Clovis dates bear this out; from east to west, north to south, the Clovis artifacts fall within that 11,000-11,500 BP slot. Nor did expansion—or technological innovation—cease. For the Clovis point gave rise to an even sleeker, more advanced projectile point, dubbed the Folsom point, after Folsom, New Mexico, where a specimen was first uncovered in association with the skeletons of extinct bison. Similar fishtail stone points have been found as far as the southernmost tip of South America—Fell's Cave in Tierra del Fuego—and dated at about 10,000 BP.

Pleistocene extinctions

The transition from the use of Clovis to Folsom points approximately 11,000 years ago coincides with the extinction of Pleistocene mammoths, horses, camels and several other varieties of megafauna—which prompted Haynes's colleague, Paul Martin, to formulate the "overkill theory," ascribing extinction to the Llano influx and insisting that they were the first inhabitants of the Americas.

Proponents of the idea of a substantial pre-Llano presence take issue with these speculations. For one, they assail the proposition that megafauna extermination stemmed directly from an invasion of Clovis. More to blame, they say, were drastic changes in climate. With deglaciation, the desert moved north, wiping out huge areas of grassland once used for foraging. They cite the work of Russell Graham of the Illinois State Museum who, having recently completed a comprehensive examination of Pleistocene fauna, concludes: "Man's pernicious effect on the modern environment is not necessarily indicative of his impact on ancient environments. . . . Undoubtedly man's predation had an effect on the megafauna, but climatic changes are the best explanation for Pleistocene extinction." Opponents of "overkill" also wonder how it is that one of the most heavily hunted of the species, a variety of bison, is still with us, while hundreds of other animal species that Clovis and Folsom did not hunt perished. Says Dennis Stanford: "Throughout life's entire history animals have gone extinct—in most cases without any help from man."

Pre-Llano proponents also have trouble accepting the presumption of Clovis's lightning-like sweep through the hemisphere. Says Stanford: "I find it impossible to accept this idea of rapid migration. Primitive cultures tend to be conservative, hunters who explore, retreat, explore, retreat. As they move from environment to environment, they must learn to adapt, and that doesn't happen overnight." MacNeish has similar reservations: "A group of primitive people traveling into completely unknown territory would have frequently taken the wrong direction, and the group would have always been saddled with household equipment and baggage, babies, pregnant women and hobbling elders.''

MacNeish proposes yet another sort of paleo-Indian-advance theory—small-group filtering or, more colloquially, the ''hurry-up-and-wait'' process: A band of migrants might be especially adapted for subsistence in broad ecological zones, he says, ''and within these zones they would be able to move rapidly, but movement from one zone to the other would require that they build up a whole new subsistence complex; that would take considerable time. The hypothesis that Clovis and Folsom moved through dozens of radically different environmental zones from the Bering Strait to Tierra del Fuego in a thousand years thus seems unreasonable.''

As such, the pre-Llano advocates suggest that Clovis, with his advanced tool kit, developed from an indigenous population in the Americas before 11,500 years ago. ''It wasn't the people that swept through America,'' says Stanford, ''but the (technology) that diffused rapidly through already existent populations—much as the idea of tobacco use traveled from the United States through Europe to the Eskimos all in a matter of a few years.''

The Siberian connection

But if paleo-Indians did indeed poke their way into the Americas long before the emergence of the Clovis Culture, who were these migrants and how did they get here, considering that a severe ice age was upon the land? Stanford: ''Soviet archaeologists have found evidence that man inhabited Siberia certainly 35,000 years ago and perhaps as early as 70,000 years ago. The discovery of early occupations of Siberia greatly increases the time available for man to come across the Bering land bridge.''

Moreover, he says, there is now reason to belive that the ice-free corridor from Alaska to North America was open for movement south for much longer periods than previously supposed. ''In fact,'' says Stanford, ''it may have been closed for only a short time during the whole (Ice Age) period. So it would have been possible for early hunters. . .to have entered North America long before 12,000 years ago and to have moved southward, continuing to exploit grassland environments.'' Further, he says, there is now even a slim possibility that an ''alternative route'' to the interior corridor may have been available—an ocean-side roadway that trailed down along the emerged Pacific coast.

Haynes has strong objections to this idea of a coastal route. ''Even if there were such a route between the glacial ice and the ocean,'' he says, ''it would have been an incredibly treacherous environment to negotiate. Under

Clovis migration? The migration of people across the Bering Strait and through North America and South America. Circles indicate Clovis and pre-Clovis sites.

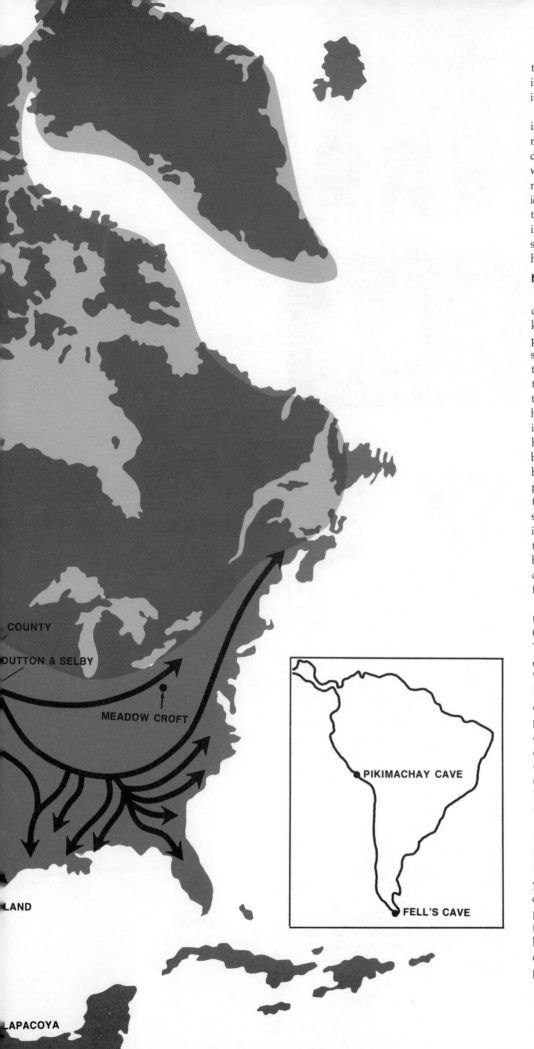

COUNTY

DUTTON & SELBY

MEADOW CROFT

LAND

LAPACOYA

VALSEQUILLO BASIN

PIKIMACHAY CAVE

FELL'S CAVE

the prevailing circumstances it's hard to imagine people moving down the coast even in boats."

By the same token, Haynes has no problem in living with the notion of Clovis's rapid migration. As he explains: "The phenomenal dispersal of Clovis sites is more compatible with a distinct migration, related to a relatively rapid, natural event—the separation of the ice sheets to form a trans-Canadian passage—than with a sudden outgrowth from meager, indigenous cultures after 12,000 years of sluggish development for which a continuity has yet to be demonstrated."

Native technology

And how does Haynes respond to the contention that Clovis's technological know-how was born in the Americas, a product of progress in home-grown artisanship and not an import? "I do not see anything that is on a developmental sequence leading to Clovis," he says. "What you see is technology akin to Eurasia, to mammoth hunters of the Old World." He sees likenesses in the tool kits of Clovis and Old World hunters, including bifacial stone scrapers, burins (chisel-like implements), flakeknives, a bone technology of bevel-based, cylindrical points and foreshafts, shaft wrenches and the use of red ocher with burials. "The similarities are unmistakable," he says, "and to invoke independent development of all these traits in the New World from a population base for which there is only tenuous evidence does not seem reasonable as does an origin from the Siberian Paleolithic."

Conpicuously absent from the Old World tool kit, however, is the centerpiece of the Clovis Culture: the fluted projectile point. This absence is probably the key piece of evidence against those who propose an Old World origin and rapid dispersal for Clovis. Haynes, in response, explains that the development of fluted, bifacial projectile points could have taken place in Alaska or along the ice-free corridor between 14,000 and 12,000 years ago. But if so, goes the counterargument, why is there no good evidence of Clovis points in this initial New World dwelling place as is the case elsewhere in the hemisphere?

The fact is, a few fluted points *have* been found in the Far North in recent years. However, one Alaskan find, said to date from the post-Clovis period, about 9,000 years ago, is held up as evidence that Clovis developed out of an already-extant American paleo-Indian culture, and that this new technology traveled not from north to south but from south to north, representing a cultural backwash. But two other fluted points, from the Putu site in the Brooks

Points and artifacts. Clockwise from the top left are: Clovis points from the Anzick Site, Montana; three Folsom points (left to right): Linger Site, Colorado, Midland Site, Texas and Meadow Ranch, Wyoming; three Vasquillo artifacts; two "artifact-looking" natural stones from the Pliocene level, Yuma County, Colorado; and three Alaska fluted points from (left to right): Driftwood Creek, Putu and Girl's Hill.

Dennis Stanford/Smithsonian Institution.

Range, have been related to a charcoal date of 11,500 years ago. "If this date is valid," says Haynes, "it makes at least two Alaskan fluted points as old as the oldest fluted points from interior North America."

Ultimately, settlement of the controversy over the early peopling of America will likely come only with the discovery of new archaeological sites. If, for the sake of argument, a site with a Clovis point were to be found in Central America, bearing a date beyond 14,000 years ago, that would all but demolish the Clovis fast-migration theory. One the other hand, one in Alaska more than 14,000 years old would help support it. But when and where new sites will turn up remains unpredictable. "Most early man sites, probably 99 percent of them, tend to be destroyed by climate almost immediately or by subsequent geologic processes," says Stanford. "When we find one, we're usually dealing with a geological freak."

On balance then: The evidence, if not altogether conclusive, certainly strongly suggests that the New World was visited and settled by migrant hunting bands from the Old World in the shadowy recesses of time back far beyond 12,000 years ago. But it remains to be determined whether the Llano Complex developed from a local, as-yet-undiscovered, indigenous progenitor or whether it originated in the Old World and spread by way of rapid dispersal. At the moment, both positions seem equally defensible. Perhaps the wisest counsel for now is for New World archaeologists to wait and see—to postpone final judgment until new, clarifying evidence comes to light. In Haynes's words: "I think that if pre-Clovis man was really here, good evidence will be found. The important thing is not to rush into it. . . . What we are actually looking for is what really happened, not what we think happened." •

The National Science Foundation contributes to the support of the research discussed in this article through its Anthropology Program.

Cultural Evolution

Patterns among human hunter-gatherers and non-human social animals suggest patterns of early hominid social development

In the long history of life on earth, *Homo sapiens* is unique—a breakthrough species. The product of several million years of natural selection and genetic mutation, human beings are capable of evolving nongenetically to a far greater extent than is any other species. They adapt biologically as other species do. But beyond that they create and transmit new environments, new organizational and behavioral adaptations and consequently new pressures for additional adaptations in a spiraling and unprecedented type of cultural evolution. They are unique in a world they never made but which, for better or worse, they are remaking.

As evidence of human cultural origins, of the roots of human social behavior, the fossil record of traditional paleontology and anthropology is unsatisfactory. It is being supplemented, however, by first-hand observations of individuals, human and nonhuman, in communities rather less complicated and hierarchical than those typical of what passes today for civilization. In many cases such observations are guided increasingly by new theoretical approaches to biology and behavior. These, in turn, may lead to the discovery of social universals, laws that cut across species lines. Such universals could help account for a wide range of behavior patterns that modern human beings appear to share with predecessor humans and protohumans, and with a host of social nonhumans as well.

To some extent, the human past is not yet dead. As long as stretches of wilderness endure there will be people who continue to subsist principally on what the land has to offer, on natural supplies of wild animals and wild plants.

What extant hunter-gatherer cultures can teach us about the remote past is limited in many respects; they are anything but relic Stone Age populations. Besides having been influenced in most cases by incursions from the outside world, these, too, are cultures that have evolved. It is generally accepted among anthropologists these days that humans and protohumans, at any stage of development, are more than way stations on the road to modern humankind. They are fully adapted creatures—adapted to the circumstances that enfold them. They are changing constantly, if they are to survive, as their heritage in the earth continually offers them new challenges. Nevertheless the fact remains that, of all extant societies, those of today's hunter-gatherers most closely resemble the

Hunter-gatherers. The San (or !Kung) Bushmen of northwestern Botswana are among the most studied of surviving hunter-gatherer societies. Their living patterns, though highly evolved, can offer clues to the social development of early *Homo*.

Melvin Konner; Richard Lee; Marjorie Shostak/Anthro-Photo

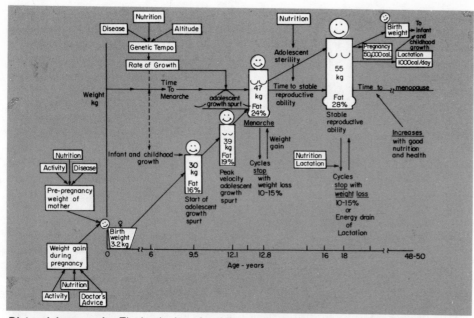

Diet and demography. The beginning of a girl's menstrual cycles and maintenance of regular menstrual function are both dependent on a minimum fat/lean ratio. Availability of food could provide a natural control of fertility among hunter-gatherers.

Rose Frisch, copyright 1978 by the American Association for the Advancement of Science

cultures that flourished everywhere that humanity flourished when it was very young.

Present-day hunter-gatherers do what survivers do best: adapt biologically and culturally to far-from-ideal conditions. It is a subtle process, calling for an intimate knowledge of the land, a symbiosis with nature and an ability to deal with the tensions and distributions of responsibility that always arise when creatures come together in groups.

There are universalities in such patterns, but for years they were not seen. Indeed, for decades during which hunter-gatherer societies were known, it might be said that hardly anyone looked.

Observing is not as straightforward as it seems. Anthropologists, like the rest of us, sometimes see only what they want or expect to see. In the case of the hunter-gatherers, the assumption had been that people at large in wildernesses, unkempt, with few possessions and exposed to the elements, must be living constantly on the edge of social disintegration.

Broad, multidisciplinary studies of Bushmen in the Kalahari Desert of southern Africa, launched in 1963 by Irven DeVore of Harvard University and Richard Lee of the University of Toronto, have done much to demolish the myth that foraging is or was a bitter and brutal struggle for survival. It was a major contravention of the conventional wisdom, for example, when Lee found that Bushmen in northwestern Botswana were not constantly harried by food

needs. They spent most of their time relaxing, telling stories and joking and visiting relatives, he reported. Some food or equipment preparation was done during such "social" time. Nevertheless, only some two to three hours a day were devoted to the food quest, even during a time of severe drought. Since these seminal observations, the Bushmen have become the most thoroughly studied of the surviving hunter-gatherer groups.

Population dynamics

Bushmen have been camping out in the Kalahari for many thousands of years, ample time to evolve sophisticated subsistence techniques. As pointed out by Henry Harpending of the University of New Mexico, where food is scattered, people must keep on the move. A single hunter-gatherer community can exploit areas of several thousand square miles. Furthermore, food is often sparse when it is found, so people go about in loosely organized bands averaging perhaps 25 to 30 persons. Individuals and families may shift from band to band depending on intragroup tensions and local food abundances. According to Harpending, more than 40 percent of Bushman parents are born outside their home-range areas, an index of outbreeding, genetic mixing and "hypermobility."

Despite the importance of controlled population growth in such societies, however, there is no need for structured population-control measures; patterns of behavior and/or biology have emerged to

take care of that problem. Bushmen do not practice sexual abstinence or contraception, and women appear to have a low rate of spontaneous abortion. Yet the interval between births is unusually long, mothers averaging one infant every four years or so. Although the phenomenon is critical to an understanding of the population dynamics of primitive societies, it remains unexplained. But suggestions are emerging from efforts to understand the reproductive biology of complex as well as simple systems.

Edwin Wilmsen of Boston University, for example, has brought back from the Kalahari analyses of blood samples collected from 29 women of child-bearing age. These analyses uncovered markedly low levels of estradiol, a sex hormone associated with fertility. Twenty-one of the women showed estradiol levels below the lowest levels measured in a control group of non-Bushman volunteers.

Two Harvard studies provide what may be additional clues to population control mechanisms. One linked to estradiol production, is affected by behavior. The other connects more broadly to fertility as it may be affected by diet.

In a series of dawn-to-dusk observations, Melvin Konner, Marjorie Shostak and Carol Worthman of Harvard found that Bushman mothers breast-feed their infants as frequently as five times an hour. Since frequent nursing stimulates high production of prolactin, another hormone and one that inhibits estradiol secretion, and since offspring are generally nursed until they are between three and four years old, that could help account for the four-year birth spacing.

And Rose Frisch of Harvard's Center for Population Studies, in a broader study of nutrition, body weight, menstruation and fertility, finds a significant relationship. Frisch and Janet W. McArthur of the Harvard University Medical School found that the beginning of a girl's menstrual cycles and the maintenance of regular menstrual function are each dependent on the attainment of a minimum weight for height, a ratio that apparently represents a critical level of fat storage. These data imply that a particular body composition of fat/lean may be an important determinant for female reproductive ability.

The stored fat, in the range of 26 to 28 percent of body weight at maturity in American and most European girls, would provide energy for a pregnancy and lactation in times of fluctuating food supplies. When fat storage drops below a certain minimum, menstruation ceases.

This is a condition Frisch and McArthur and others have noted in otherwise normal women, such as ballet dancers and professional runners, who restrict their food intake and/or exercise excessively.

The cessation of menstrual cycles is also observed in girls and women with self-imposed starvation accompanying the psychogenic disease of anorexia nervosa. Frisch has also reported, from studies of historical populations, that slow growth to maturity due to less favorable nutrition is associated with a shortened, less efficient reproductive span.

Data from experiments done by Frisch, D. M. Hegsted and K. Yoshinaga, in which calories from fat were substituted for the same number of calories eaten as carbohydrate, suggest that high fat diets may contribute to earlier sexual maturation in human as well as in animal populations.

Dietary data gathered by Wilmsen suggest a relationship between Frisch's data and conditions among Bushmen: The average adult weights among Bushmen vary from some 46 kilograms in June, July and August, after the rainy season when food is relatively plentiful, to about 43.5 kilograms in the December-February period following the dry season. The relationship to fertility is not established, but can be inferred.

Sociability

Not only biological, but behavioral changes, too, seem to be imposed by the requirements of group living. There is no life style as intimate—and hence demanding—as that of the hunter-gatherer. According to Patricia Draper of the University of New Mexico, Bushmen cluster around Kalahari campfires "almost like a flock of birds," surrounded by bush country "stretching away for miles in every direction...vast, undifferentiated, unhumanized." Under such conditions, huddled together in an immense wilderness, intragroup harmony is an essential of group, hence individual, survival.

Limited food resources and small band size, Draper notes, help keep tensions reasonably low. When a Bushman child quarrels with a playmate, aggression rarely has a chance to build. Someone, not necessarily a parent, is always there to soothe ruffled feelings and separate the combatants. From the age of six months, children are trained to share and give presents.

Throughout life the accent is on cooperation and quality. Leadership tends to pass from individual to individual. The best hunter naturally plays a central role

The fierce people. Children of the settled but warlike Yanomamo celebrate an important cultural activity that more fragile hunter-gatherer communities can apparently ill afford.

Napoleon Chagnon/Anthro-Photo

in planning hunts; the best storyteller entertains around fires at night; the best healer is consulted in times of sickness. But there are no official, full-time leaders.

Similar coping patterns exist among aborigines in the deserts of western Australia. They also keep group sizes down, living in bands of about 25 to 30. They separate when tensions mount and have special rituals to minimize mayhem when trouble starts. In a minor quarrel, say over the sharing of food, the antagonists may stand some hundred yards apart, hurl boomerangs at each other in turn (hard, but not too accurately), call each other names and then rush fiercely toward one another as if intent on murder. Instead they hug and weep and become friends again. For more serious offenses, such as wife-stealing, the ritual may start the same way. But instead of hugging, the men settle things by falling to their knees, close together; the offender receives a shallow knife cut on his back.

The aborigines have adapted to conditions at least as arid as those prevailing in the Kalahari. The food quest may send them on treks of more than 1,500 kilometers, and survival depends above all on knowing where water is, on firmly imprinted mental maps and the ability to detect landmarks in the most barren terrain. Richard Gould of the University of Hawaii, who has lived among aborigines and excavated in Australia's Western Desert, estimates that the average adult remembers the locations of

more than 400 sources of water, everything from the few permanent water holes to small pools in shaded rock clefts and hollows in trees.

Shift to farming

Complex relationships between diet and fertility must also have been important in times past. During two million years of human hunting and gathering, it is inferred from fossil records that local populations increased very slowly if at all. Our ancestors' numbers began soaring with a vengeance, however, after they shifted from foraging to farming, perhaps because of the fertility-raising effects of milk, cereal grains and other domesticated foods.

Something similar seems to be happening today in Botswana. Bushmen settling in villages are having more children than do their wide-ranging relatives. Their birth spacing is down from four years to two or less. Village Bushman societies are also less harmonious. Tensions are less easily controlled, quarrels are more numerous and the egalitarianism that characterizes the hunter-gatherer society appears also to be breaking down. Relative abundance may be the key to the behavior shift.

As far as the availability of food is concerned, it would be difficult to find a part of the world less like the Kalahari than that occupied by the Yanomamo tribe of southern Venezuela and northern Brazil. The Yanomamo live in dense jungle where resources are abundant and evenly distributed rather than scattered. Groups—villages, in this case—are relatively large and close enough to each other for commerce of a kind to have emerged.

Napoleon Chagnon of Pennsylvania State University, who has been studying the Yanomamo ever since his graduate-school days in 1964, calls them the "fierce people," and with good reason. The world they live in is essentially a uniform world. Not only is food evenly distributed, but so are clay for pottery, strong vines for hammocks and bamboo for arrowheads. So, too, are human resources; every village includes individuals skilled in the making of pottery, hammocks and arrowheads.

Yet the Yanomamo live in a state of chronic warfare. Their consuming passion is ambushes and sneak attacks, often at dawn. A village may raid or be raided half a dozen times a year. One out of every four adult males dies in battle. Nevertheless, fertility is high; populations have been on the rise for at least 200 years.

The Yanomamo seem to be a driven people. They recognize that fighting is bad and want to stop, or at least say they do, Chagnon reports. But, under social pressures yet to be defined, they cannot control behavior which to us appears irrational. The real world as it would be described by any objective observer is not the Yanomamo's world. They live in a different jungle, a jungle they have created culturally.

Distinctions by gender

There appears to be a population density and life style concomitant as well to equality between the sexes. As far as Draper could discern, adults among the hunter-gatherer Bushmen do not discriminate by sex in dealing with children, although she did observe some differences in the behavior of boys and girls.

Fundamentally, girls tend to remain closer to the center of the camp and to adults, while boys tend to go off by themselves. Such differences may not make much of a difference among nomadic Bushmen who have time to burn. But Bushmen living in agricultural villages have far less leisure, and that seems to reinforce differences between the sexes. Draper notes that girls, being more often near home base and within earshot, are naturally called on more often to do household chores. Boys, who prefer—and are encouraged—to range more widely, are often found away from

the village center and are assigned such tasks as tending herds.

The powerful impact of village life, however, has less to do with work itself than with attitudes toward the work. Sexual division of labor is nothing new. In foraging bands women traditionally do most of the gathering and men do most of the hunting. But men often gather and women may hunt upon occasion. It is only within a village context that Bushmen begin to look on women's work as inferior and make permanent distinctions among assigned roles.

A great deal remains to be learned about human adaptation, cultural and biological, not only from the Bushmen and Yanomamo but also from Australian aborigines, Eskimos and representatives of other vanishing cultures. For the hunting-gathering band certain features, such as sharing and small group size, seem to be typical of all societies studied to date and probably reflect the traditions of prehistoric times. But sweeping, all-or-nothing statements are exceedingly risky. About the only safe statement is that humans are adapters supreme. "The important thing," says Sherwood Washburn of the University of California at Berkeley, "isn't hunting and gathering, but the ability to shift, from an all-meat to an all-vegetable diet and back again if necessary. Flexibility is the essence of being human." It may well have been the adaptation that made all other adaptations possible.

Other species

While some scientists tackle problems of human origins head-on, investigating *Homo sapiens* directly, others contribute findings from observations of other living species. These creatures are studied for themselves, for their richly varied survival strategies and as living examples of the biological matrix out of which humankind emerged. They are also studied because, although none of them are simple, they are all much simpler than ourselves. By contrast and comparison, they accentuate the magnitude of humanity's own prodigious complexity.

The range of adaptations available to the members of a given species is impressive but limited. For all the varieties of communal living, for all the flocks and herds and prides and troops, behaviors are known that appear to contribute to individual survival, and they are easy to understand. But there are other patterns, through which individuals contribute to group survival often at their own expense. To explain such behavior, it is

increasingly being proposed that behavioral as well as biological attributes may be explained by natural selection.

Some investigators have reservations about this way of looking at things, particularly as it is applied to higher orders of animals. But it seems to help account for an increasing body of observation. A number of major research projects are addressing the key question of how creatures programmed for individual survival can function as members of a society, and most of the projects make use of theoretical concepts developed by William Hamilton of the University of Michigan and Robert Trivers of the University of California at Santa Cruz.

The answer to the question relies on a calculus of socially productive selfishness, a way of figuring out when it profits one individual to take a risk in behalf of another, and the notion that such an equation is genetically programmed. It appears to many biologists to be a calculus based on kinship.

The British biologist J. B. S. Haldane once said he would give his life for eight cousins. An Arab proverb puts it less quantitatively but rather more eloquently: "I against my brother; I and my brother against my cousin; I and my brother and my cousin against the world."

Such a calculus predicts that an animal will be less likely to defend or cooperate with an individual with which it shares only a quarter of its genes (an uncle, aunt, nephew, niece or grandparent) than with one having half its genes (brother, sister, parent, offspring), and still less likely to make a sacrifice for those in the one-eighth-gene category (first cousins, great-grandchildren). The payoff is measured in terms of genes saved and passed along to one's own or one's relatives' descendants.

Testing kin selection

The problem is to learn to what extent these ideas apply at different levels of biological and social complexity. They were first put to the test among ants and other social insects. According to Edward Wilson of Harvard, insects "rank among the ecologically dominant animals of the land." They have evolved a number of unusual genetic relationships during the past 100 million years.

For example, in an ant colony the queen's eggs can grow into mature offspring whether they are fertilized or not. All eggs which are fertilized (diploid) develop into females. Unfertilized eggs develop anyway and become (haploid) males.

Cooperative brooders. Four adult white-fronted bee-eaters and (center) a recent fledgling. Natalie Demong examines bee-eaters' cliffside homes. Diagram tracks individual members of the bee-eater community as they either breed their own young or help other pairs raise theirs.

Stephen T. Emlen

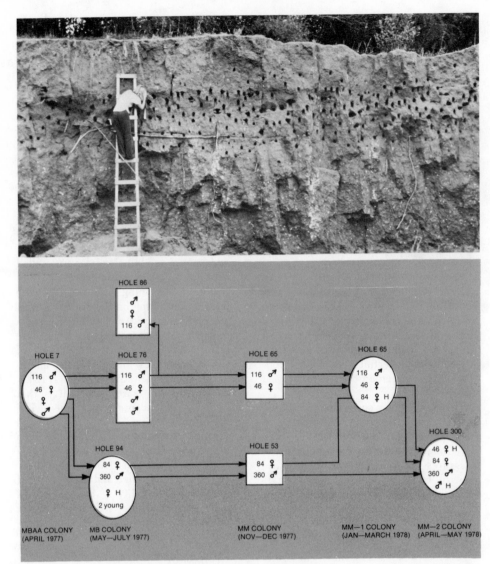

As a consequence, sisters share more genes with each other than with their brothers, than their brothers with each other, or even than the sisters do with their own offspring. This means that, if behavior has a genetic component, they should behave more altruistically toward one another than toward brothers and offspring. And they do. Such analysis according to kin-selection principles is offered to explain why males contribute little to the labor of the colony and why female soldiers fight to the death in defending the colony against invaders.

Nursemaid birds

The same principles often hold for birds, among the most monogamous of species. Voracious nestlings, all beak and beak open all the time, demand the constant attention of two busy parents, and often of others as well. Cooperative breeders or "helpers," birds that forego breeding themselves and participate in the rearing of other birds' offspring, have been observed in more than 150 species,

including the white-fronted bee-eater currently being studied in Kenya's Lake Nakuru National Park by Stephen Emlen and Natalie Demong of Cornell University.

These robin-sized birds dig burrows up to six feet deep into cliffs and river banks, live together in groups of two to seven and depend on the good services of helpers, which may make up 20 to 50 percent of all adults. Cooperation pays off; mated pairs with two helpers rear more than twice as many offspring as do unassisted pairs. Finding out whether helpers help kin preferentially, however, demands more intensive studies. Every individual must be identified as to its relationships to other individuals. To obtain complete genealogies, some 800 bee-eaters have been marked with leg rings and plastic wingtags.

Nakuru bee-eaters live in a harsh, unpredictable world. One mated pair (tagged F46 M116) lost its young last year in an April flood, spent the rest of that

breeding season living with another pair in another colony, and then moved to still another colony. There the pair remated, was joined by a female helper, and nevertheless lost a second brood in a second flood. This time the birds separated to help pairs in different nests, only to reunite later for a third family-rearing effort.

Although a great many moves remain to be analyzed within and among colonies numbering 50 to 500 individuals, Emlen believes that kinship helps determine who helps whom. "But whatever the nature of such bonds," he states, "it is the faithfulness of their reestablishment that gives stability and pattern to the fabric of bee-eater society."

Mammals

Tracing the family trees of mammals is even more complicated, mainly because monogamy is rarer. Polygamy tends to occur when conditions give one sex an appreciably higher opportunity for multi-

Danger! A female Belding's ground squirrel sounds the alarm at the approach of danger. Kinship to nearby squirrels appears to influence warning behavior.

George D. Lepp

ple mating. This is a situation which, in mammals, is usually though not always exploited by the male; the task of nourishing the young falls disproportionately on the female.

One investigator concerned with mammalian behavior is Paul Sherman of the University of California at Berkeley, whose subjects include some 2,500 Belding's ground squirrels, each marked with ear tags and black hair dye, in a 20-acre Sierra Nevada meadow. His observations apply to females only; males do not stay around the burrow after mating. Ground squirrels emit a sharp staccato call at the approach of weasels, coyotes and other terrestrial predators, and females with close relatives nearby (sister, daughter, mother) sound the alarm more often and sooner than do those without close relatives. Sherman believes that these ground squirrels may recognize close relatives on the basis of smell or sight patterns imprinted on their memories at an early age. Cousins, nieces and granddaughters are in effect strangers; their presence elicits no more alarm calls than does that of nonrelatives.

The subtleties of individual interactions increase enormously among primates. In most group-living primates, females form the stable core of the troop while males leave at adolescence and seek places in other troops. Exceptions exist to practically every rule of animal behavior, however, and often they demand special study because they may lead to productive surprises. For example, a challenge as yet unmet is to explain why chimpanzees, humankind's closest living relatives, are one of the few social primates in which females rather than males may move away from their home troops.

These apes are known best from the widely publicized work of Jane Goodall who spent more than 15 years with them in the Gombe Stream Research Center in Tanzania. Ultimately, they accepted her almost as an honorary member of the troop, if not quite as a fellow chimpanzee. And she was able to observe many strikingly human activities among them: chases and tug-of-war games among juveniles; old friends kissing and embracing; "rain dance" rituals during storms; tools being made out of grass stalks and vines broken to just the right length for extracting succulent termites from their nests; a male waiting for a companion, presumably a female, and fidgeting "like an impatient man looking at his wristwatch."

Chimps' behavior reminiscent of humans' may not be too surprising in a species with which humans as a species share more than 99 percent of their genes. But it goes deeper than that. There are long-lasting family ties. One female who was probably more than 50 years old had two sons, both over 20, who often groomed her and roamed with her through the Gombe forest. Brothers may maintain their status and defend one another when threatened by a higher ranking member of the male hierarchy. Wider associations are also formed: clans of a kind, closed social networks of 20 to 80 individuals.

Chimpanzees eat meat, a rare custom among nonhuman primates. Even rarer, they may hunt cooperatively, stalking their prey, often a monkey or small antelope, silently and maneuvering to cut off escape routes. Relationships among troop members change dramatically when an animal is killed. The killer may occupy a lowly position in the hierarchy under normal conditions, but after the kill he is ape *uno*.

Geza Teleki of Pennsylvania State University has worked with Goodall and reports on 167 "predatory episodes"; he describes how other troop members defer to the killer, holding out their hands palm up in the classic begging gesture for a scrap of meat. Sharing occurs, but not ordinarily and often reluctantly. About a third of all requests are honored; whether kinship is a factor can as yet only be guessed.

Carnivores in packs

The general impression from observing chimpanzees in the wild is how near they are to people, and yet how far. Activities that evolved to a high point in humans remain undeveloped in them. The problem is to deduce what changes, environmental and otherwise, would make such activities necessary for survival—and, in the process, transform a chimpanzee-like creature into a hominid, a member of the family of man.

One possibility involves the coming of drier times, the dwindling of forests and forest foods, and the spread of open woodlands and grassy savannas. Early hominids may have adapted to such conditions by becoming regular carnivores—and by needing each other. According to Washburn, fossil counts suggest that there was only one hominid

Complex behavior. Chimpanzees (left) share the carcass of a monkey and (above) use grass-stem tools to fish for termites in a mound.

Geza Teleki

on the African savanna for every 50 to 100 baboons, roughly the expected proportion of predators to species that do not depend on predation. "Meat was the limiting factor," says Washburn. "Meat became especially important to carry hominids through the dry season, when it made up a large part of the diet."

To appreciate the full socializing effects of meat-eating, there is at least as much to learn about human origins from social carnivores as from our fellow primates. Another denizen of the African savanna, the wild dog, has developed social living to a fine point. "The most striking aspect of African wild dog sociality," according to George Schaller of the New York Zoological Society, "is the amity that exists between members....I never saw two dogs fight. In competition for a bone or in other situations in which a fight could erupt, both animals tend to assume the appeasement posture (body lowered, ears flattened), thereby terminating the interaction. Packs seem to lack a rigid hierarchy."

Similar observations by Richard Estes of the Philadelphia Academy of Natural Sciences and Wolfdietrich Kuhme of Germany confirm the general picture of controlled aggression and cooperation. A typical hunt may start quite innocently, almost casually, with the pack at ease lolling about in the grass. Suddenly one dog gets up and moves playfully toward a neighbor, and the two begin romping and chasing one another. Then the pace quickens, as other dogs join in. Gradually the intensity mounts, building up to a wild crescendo, the entire pack milling about, circling in a kind of organic eddy, emitting peculiar, twittering, birdlike calls. This

Social insects. Leafcutter ants in their fungus garden. Gardener-workers are dwarfed by their queen. Complex organization and genetically determined behavior may both be at work.

Carl W. Rettenmeyer

Social carnivores. Members of a pack of African wild dogs hunt, kill and feed cooperatively and bring food home to young and others left behind.

George Schaller

ritual, a canine pep rally, brings the pack to a fever pitch.

Perhaps a dozen adults, males and females, are then ready for a kill. They form a highly effective hunting team, making use of stalking, sudden rushes and surrounding tactics when a victim has been selected. Wild dogs on the run can average 30 miles an hour for several miles, about 85 percent of all chases ending in a kill. They eat some of the meat on the spot and carry the rest away in the forms of chunks swallowed whole, to be regurgitated back at the den for pups and their baby-sitting guardians. The wild dog's life style represents an outstanding example of stability, the result of a strategic trade-off between the needs of the individual and the needs of the group.

These are boom times for research on the evolution of social behavior. It can be seen in the pace of research and in the enthusiasms of committed scientists. Many are engaged in an accelerating search for general rules, such principles as kin selection, which could affect behavior in many social species, from insects on up. There is a new respect for animals, which are studied increasingly in their own right as successful adaptations, and not as mere stages on the way to *Homo sapiens*.

Work in progress indicates the shape of future advances. Wilson is in the midst of new studies of the societal structure among leaf-cutter ants. These ant societies include specialized foragers, soldiers, "baby-sitters" to take care of the young and "gardeners" to grow yeasts and fungi, even providing a formic "manure" consisting of fecal droplets. (One colony of some 10,000 workers is busy cutting up the leaves of a plant housed in Wilson's home outside Boston.)

Emlen's studies of bee-eaters have several years to go. Primate projects include research by Sarah Hrdy of Harvard on conflict and common causes among langur monkeys, Alison Richard of Yale on how group living pays off for rhesus monkeys and Richard Wrangham of Cambridge University in England on what determines whether the permanent core group of a troop will consist of males or females. Trivers, who has done much to provide the theoretical framework for these and other sociobiological investigations, is probing deeper into the dynamics of social behavior in studies of wasps, other social insects, African hunting dogs and deer.

Creative self-interest

Several years ago Trivers published a seminal paper on conflict between parents and offspring, the former prepared to invest a certain amount of time and energy in rearing the young, the latter actively competing for all the attention they can get (which is generally rather less than they want). He presents the infant as "a psychologically sophisticated organism," surprisingly sensitive to adult behavior and determined to influence that behavior in its own behalf. Trivers's ideas apply to sexually reproducing species from birds and rats to caribou and primates, human as well as nonhuman.

Now he is working toward a still broader theory of the family which, among other things, goes into tensions between father and daughter and mother and son. It is not a Freudian theory: "The source of conflict is not repressed mating urges, but differences in genetic self-interest." In other words, Trivers suggests, the conflict is among individuals competing and manipulating others primarily to ensure that their own genes will be transmitted to future generations.

Self-interest is also heavily involved in choosing mates, although not necessarily in accordance with the notion that the strongest males compete for the most attractive female, with the winner getting her. The odds are that the female's role has been greatly underrated. The antlers of a male deer, for example, may be designed less to overawe or defeat rival male suitors than to inform the female as to his physical condition and temperamental suitability to sire her offspring.

Further analysis is called for, and Trivers warns that "all this may be dead wrong." Whatever the outcome, such thinking plays a major role in identifying new problems and questions to ask, stressing the need for new research and a harder look at past findings. Behind all the experiments and theorizing is a drive to identify more precisely the basic similarities and the basic differences between humans and other species.

A recent statement by Pierre van den Berghe and David Barash of the University of Washington summarizes the situation today: "A century after Darwin, we have learned enough biology to try to apply it to behavior in general, social behavior in particular and human social behavior most especially." They might additionally have said: and to the question of what, in the evolution of culture, helped to form human beings and to make them what they are today.•

The National Science Foundation contributes to the support of research discussed in this article through its Anthropology and Psychobiology Programs.

GLOSSARY

aborigine. One of the original or earliest known inhabitants of a country.

anthropoid. Resembling a human being; applied most often to the most highly developed apes.

arboreal. Living in or among trees.

Australopithecus. Sometimes called "man-apes," the hominids in this genus made and used simple stone tools.

bipedal. Having two feet; two-footed.

brethren. The fellow members of a society.

denizen. An inhabitant.

dentition. The growth and arrangement of teeth.

egalitarianism. A principle based on ideas of political and social equality.

eland. A large, heavily built African antelope with twisted horns.

estrous. Of or having the characteristics of estrus, the sexual excitement, or heat, of female mammals.

gracile. Slender; thin.

guano. A substance composed chiefly of the dung of sea birds or bats, accumlated along certain coastal areas or in caves.

hominid. A family of two-legged primates that includes all forms of humans, extinct and living.

hominoid. Of the form of or resembling a human.

Homo erectus. The advanced tool and culture maker that was the ancestor of all later forms of human beings. From about 1.5 million years ago until about 400,000 years ago, *Homo erectus* was the only human species on the planet.

Homo sapiens. Humans, having first appeared some 400,000 to 500,000 years ago.

Homo sapiens sapiens. Modern humans, having first appeared about 50,000 years ago.

indigenous. Originating in the region or country where found.

mandible. The jaw. In vertebrates, the lower jaw.

masticatory. Having to do with chewing; adapted for chewing.

mastodon. A very large, extinct mammal, much like an elephant.

morphology. The branch of biology dealing with the form and structure of animals and plants without regard to function.

morphometry. The measurement of the external form of any object.

Neanderthal. An extinct, primitive human race, widespread in Europe, North Africa, and western and central Asia in the early Stone Age. The first fossils were discovered in 1856 at the Neanderthal Valley near Dusseldorf, Germany. The Neanderthal of Europe had a large, heavy skull and low forehead; a broad, flat nose; and a heavy lower jaw with teeth intermediate in shape between those of modern humans and those of the apes. He averaged five feet in height, and is believed to have lived between 150,000 and 30,000 B.C. *Also called* Homo sapiens neanderthalensis.

paleoanthropology. The study of the earliest human races as represented by their fossils and the remains of their cultures.

paleoclimatology. The study of the climate in earlier geological times.

paleoecology. The study of the relationship of living things to their environment and to each other in prehistoric times.

palynology. The study of plant spores and pollen, especially in fossil form.

Pleistocene. The geological epoch before the present period, characterized by the rise and recession of continental ice sheets and by the appearance of humankind.

pongids. The anthropoid apes, including the chimpanzee, most closely related to humans.

primates. Having its origin some 60 million years ago, this is the order of the most highly developed mammals, including, but not restricted to, the apes and humans.

primeval. Of or having to do with the first age or ages.

Ramapithecus. Sometimes considered to be the first stage in human evolution, this small-brained creature stood about 3 feet tall and weighed approximately 40 pounds. Fossils indicate that *Ramapithecus* dates back about 14 million years, and became extinct about 8 million years ago. Specimens have been unearthed in Greece, Turkey, Hungary, Africa, and India.

simian. 1. Like or characteristic of an ape or monkey. **2.** An ape or monkey.

tectonic. Of or resulting from changes in the structure of the earth's crust.

unicuspidate. Having only one cusp, as an incisor or canine tooth.

INDEX